# FET MODELING FOR CIRCUIT SIMULATION

**THE KLUWER INTERNATIONAL SERIES**
**IN ENGINEERING AND COMPUTER SCIENCE**

VLSI, COMPUTER ARCHITECTURE AND
DIGITAL SIGNAL PROCESSING

*Consulting Editor*

Jonathan Allen

**Other books in the series:**

*Logic Minimization Algorithms for VLSI Synthesis.* R.K. Brayton, G.D. Hachtel,
  C.T. McMullen, and A.L. Sangiovanni-Vincentelli. ISBN 0-89838-164-9.
*Adaptive Filters: Structures, Algorithms, and Applications.* M.L. Honig and
  D.G. Messerschmitt. ISBN 0-89838-163-0.
*Computer-Aided Design and VLSI Device Development.* K.M. Cham, S.-Y. Oh,
  D. Chin and J.L. Moll. ISBN 0-89838-204-1.
*Introduction to VLSI Silicon Devices: Physics, Technology and Characterization.*
  B. El-Kareh and R.J. Bombard. ISBN 0-89838-210-6.
*Latchup in CMOS Technology: The Problem and Its Cure.* R.R. Troutman.
  ISBN 0-89838-215-7.
*Digital CMOS Circuit Design.* M. Annaratone. ISBN 0-89838-224-6.
*The Bounding Approach to VLSI Circuit Simulation.* C.A. Zukowski.
  ISBN 0-89838-176-2.
*Multi-Level Simulation for VLSI Design.* D.D. Hill and D.R. Coelho.
  ISBN 0-89838-184-3.
*Relaxation Techniques for the Simulation of VLSI Circuits.* J. White and
  A. Sangiovanni-Vincentelli. ISBN 0-89838-186-X.
*VLSI CAD Tools and Applications.* W. Fichtner and M. Morf, editors.
  ISBN 0-89838-193-2.
*A VLSI Architecture for Concurrent Data Structures.* W.J. Dally.
  ISBN 0-89838-235-1.
*Yield Simulation for Integrated Circuits.* D.M.H. Walker. ISBN 0-89838-244-0.
*VLSI Specification, Verification and Synthesis.* G. Birtwistle and
  P.A. Subrahmanyam. ISBN 0-89838-246-7.
*Fundamentals of Computer-Aided Circuit Simulation.* W.J. McCalla.
  ISBN 0-89838-248-3.
*Serial Data Computation.* S.G. Smith and P.B. Denyer. ISBN 0-89838-253-X.
*Allophonic Variation in Speech Recognition.* K.W. Church.
  ISBN 0-89838-250-5.
*Simulated Annealing for VLSI Design.* D.F Wong, H.W. Leong, and C.L. Liu.
  ISBN 0-89838-256-4.
*Polycrystalline Silicon for Integrated Circuit Applications.* T. Kamins.
  ISBN 0-89838-259-9.

# FET MODELING FOR CIRCUIT SIMULATION

**Dileep Divekar**
Valid Logic Systems, Inc.

**KLUWER ACADEMIC PUBLISHERS**
Boston/Dordrecht/Lancaster

**Distributors for North America:**
Kluwer Academic Publishers
101 Philip Drive
Assinippi Park
Norwell, Massachusetts 02061, USA

**Distributors for the UK and Ireland:**
Kluwer Academic Publishers
MTP Press Limited
Falcon House, Queen Square
Lancaster LA1 1RN, UNITED KINGDOM

**Distributors for all other countries:**
Kluwer Academic Publishers Group
Distribution Centre
Post Office Box 322
3300 AH Dordrecht, THE NETHERLANDS

**Library of Congress Cataloging-in-Publication Data**

Divekar, Dileep.
  FET modeling for circuit simulation.

  (The Kluwer international series in engineering
and computer science ; SECS 48)
  Includes index.
  1. Field-effect transistors—Computer simulation.
2. Field-effect transistors—Mathematical models.
I. Title.  II. Series.
TK7871.95.D58  1988      621.3815 '284      87-37848
ISBN 0-89838-264-5

Printed in the United States of America

To my family

# CONTENTS

# PREFACE

Circuit simulation is widely used for the design of circuits, both discrete and integrated. Device modeling is an important aspect of circuit simulation since it is the link between the physical device and the simulated device. Currently available circuit simulation programs provide a variety of built-in models. Many circuit designers use these built-in models whereas some incorporate new models in the circuit simulation programs. Understanding device modeling with particular emphasis on circuit simulation will be helpful in utilizing the built-in models more efficiently as well as in implementing new models. SPICE is used as a vehicle since it is the most widely used circuit simulation program. However, some issues are addressed which are not directly applicable to SPICE but are applicable to circuit simulation in general. These discussions are useful for modifying SPICE and for understanding other simulation programs. The generic version 2G.6 is used as a reference for SPICE, although numerous different versions exist with different modifications. This book describes field effect transistor models commonly used in a variety of circuit simulation programs. Understanding of the basic device physics and some familiarity with device modeling is assumed. Derivation of the model equations is not included.

( SPICE is a circuit simulation program available from EECS Industrial Support Office, 461 Cory Hall, University of California, Berkeley, CA 94720. )

# Acknowledgements

I wish to express my gratitude to Valid Logic Systems, Inc. for their support. This book is based on the work of many researchers and their contributions are gratefully acknowledged.

The support and patience of my wife Shubhada were of great help.

# FET MODELING FOR CIRCUIT SIMULATION

"Everything should be made as simple as possible, but not any simpler."

**- Albert Einstein**

"A model of a physical device is a mathematical entity with precise laws relating its variables. The model is always distinct from the physical device, though its behavior ordinarily approximates that of the physical device represented. Thus a model is never strictly equivalent to the device it represents."

**- John Linvill**

# CHAPTER 1

# CIRCUIT SIMULATION

## 1.1. Introduction

A simulator implements test instrumentation in software. With simulation, a designer troubleshoots a design by probing various circuit nodes as if checking out a physical board or chip with an oscilloscope. In other words, the design is analyzed and verified by displaying waveform and timing information - but without having to build the circuit. Designers who want to reduce prototyping time and shorten design cycles should be simulating their circuits. Besides the obvious advantage of fewer prototype cycles and greatly reduced development costs, designs that have undergone simulation are typically of higher quality. In other words, simulation makes it easier to optimize a design. And when the design is optimized, the transition to manufacturing is much smoother.

Circuit simulation enables the circuit designer to do

things which are otherwise not possible [1]. It makes it possible to:

-       probe nodes which may not be accessible in an actual circuit.

-       observe waveforms and frequency responses without loading the circuit.

-       avoid parasitics introduced by the breadboard but which will not be present when the actual circuit is fabricated.

-       to use device models which represent the integrated circuit devices as opposed to their discrete approximations.

-       do sensitivity, worst case and statistical analyses.

-       do circuit optimization.

-       gain better understanding of the circuit behavior by

        using ideal devices selectively to isolate the effects of various device parameters.

        feeding    ideal    waveforms,    such    as extremely fast pulses, into the circuit.

        separating dc, ac and transient behavior.

        opening feedback loops without disturbing other circuit bevavior.

## 1.2. Circuit Simulation Programs

Although circuit simulation programs differ considerably in size and capability, the structure of most simulation programs is similar [2] [3]. Most general purpose simulation programs are made up of four main stages as shown schematically in Figure 1.1.

In the input stage, the program receives information from the user with regard to the network configuration, element characteristics and types of analyses to be performed.

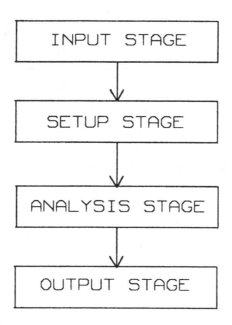

**Figure 1.1:**    Main stages of a simulation program.

The input stage reads the input information, constructs input data structures and checks these data structures for obvious user errors. After the input stage, the input data structure contains a complete and selfconsistent description of the circuit. On the average, in any user oriented simulation program, about 40 per cent of the program code is used to process the input and to provide extensive diagnostic messages. Although most of these features are conveniences rather than the essentials of the simulation program, if any simulation program is to gain wide acceptance, especially among engineers, user convenience is among the topmost factors to be considered.

After the input stage has executed successfully, additional data structures are constructed in the setup stage.

These data structures depend on the types of analyses to be performed and are used during the analysis stage.

The analysis stage is the major part of the simulation program. This stage performs the circuit analyses requested in the input stage. Although the input and output stages determine how easy the simulation program is to use, the analysis stage plays a dominant role in determining the efficiency of the program. Moreover, the algorithms that are employed in the analysis stage are much more complex than the algorithms in the other stages of the simulation program. For these reasons, the largest portion of the development effort that is expended on a simulation program is devoted to the implementation of reliable and efficient circuit analysis techniques.

Last, but not the least important in a simulation program, is the output stage, for this is where the user obtains the answers. A variety of output capabilities may be provided by the output stage and the flexibility and ease of use of this stage is an important aspect of the simulation program.

## 1.3. Circuit Analysis

The analysis portion of a circuit simulation program determines the numerical solution of a mathematical representation of the circuit [2] - [5]. The transition from the physical circuit to a mathematical system of equations is accomplished by representing each element in the circuit by a mathematical model. Thus the device models are the link between the actual circuit and its mathematical representation for circuit analysis. Device models determine the degree to which the simulation results will match the actual circuit performance.

The system of equations that describes the complete circuit is determined by the model equations for each element and the topological constraints that are determined by the interconnection of elements. Any lumped network is governed by three types of constraint equations arising from the Kirchoff voltage law, Kirchoff current law and element characteristics. These constraints constitute a system of equations. Some of these equations are nonlinear algebraic equations and the remaining are nonlinear differential equations. The system of equations is of the form:

$$f\left(x, \dot{x}, t\right) = 0 \qquad (1.1)$$

where $x$ is the vector of unknown circuit variables, $\dot{x}$ is the time derivative of $x$, $t$ is the time and $f$ is, in general, a nonlinear operator.

Circuit analysis determines the solution of these equations numerically as opposed to analytical solutions. The nonlinear differential equations are converted into nonlinear algebraic equations by using numerical integration methods. The system of nonlinear algebraic equations is then solved using the Newton-Raphson method.

### 1.3.1. DC Analysis

The dc solution is determined for the equilibrium case when the vector $\dot{x}$ is zero. Three methods are widely used in the computer simulation programs for formulating the circuit equations: (1) the nodal method, (2) the hybrid method, and (3) the state-variable method. Modified nodal analysis is more commonly used. With nodal analysis, the node voltages are the unknown circuit variables. This choice of variables automatically insures that the circuit solution satisfies the Kirchoff voltage equations. The circuit equations are then determined by writing an equation for Kirchoff current law at each node. In general, the nodal equations are of the form:

$$f\left(v\right) = i \qquad (1.2)$$

where $v$ is the vector of unknown node voltages, $i$ is the current excitation vector and $f$ is the nonlinear function representing the element characteristics. The interpretation of (1.2) is that each equation represents a summing of current contributions at each node. It should be noted that the nonlinearities are restricted to be voltage rather than current dependent. Fortunately, it is possible to formulate the nonlinear equations of the desired circuit elements as voltage controlled equations and the above restriction is usually not severe.

The solution to the nonlinear system of equations is obtained by an iterative sequence of linearized solutions. The Newton-Raphson algorithm is the most common method of linearization. Given a set of nonlinear equations of the form (1.2), for the dc case, the equations may be expressed as:

$$g\left( v \right) = 0 \qquad (1.3)$$

The nonlinear equations are transformed into a system of linear equations by applying Newton's method to the system (1.3):

$$\frac{\partial g}{\partial v} v^{n+1} = - g\left( v^n \right) + \frac{\partial g}{\partial v} v^n \qquad (1.4)$$

where n is the index of iteration. Usually, (1.4) is written in the form of a linear system of equations:

$$J v = b \qquad (1.5)$$

where $J$ is the Jacobian:

$$J = \frac{\partial g}{\partial v} \qquad (1.6)$$

The linear system of equations (1.5) is usually solved using the method of Gaussian elimination or LU factorization.

The Jacobian consists of the nodal conductance matrix of the linear elements of the circuit together with the linearized conductances associated with each nonlinear circuit element. The vector on the right hand side of (1.5) consists of independent source currents and the Norton equivalent source currents associated with each nonlinear circuit element. Thus at each iteration in a nodal analysis, the linearized conductances and Norton equivalent source currents must be recomputed and the linearized nodal conductance equations reassembled.

### 1.3.2. Transient Analysis

Transient analysis determines the time domain response of the circuit over a specified time interval $(0, T)$. The initial time point is arbitrarily defined as time zero. The initial solution is either specified by the user, or, more conveniently, is determined by a dc operating point analysis.

The transient solution is determined computationally by dividing the time interval $(0, T)$ into discrete time points $(0, t_1, t_2, \cdots, T)$. At each time point, a numerical integration algorithm is employed to transform the differential model equations of energy storage elements into equivalent algebraic equations. After this transformation, the solution is determined iteratively in the same fashion as the nonlinear dc operating point. Various numerical integration methods are available, but the trapezoidal integration method is most commonly used in the circuit simulation programs.

For transient analysis, equations (1.2) can be written as:

$$f\left(v(t)\right) = i = \frac{\partial q\left(v(t)\right)}{\partial t} \qquad (1.7)$$

This equation can be written in the integral form as:

$$\int_{t_n}^{t_{n+1}} f\left(v(t)\right) dt = \int_{t_n}^{t_{n+1}} dq \qquad (1.8)$$

Applying the trapezoidal integration formula to the left hand side yields:

$$\frac{h_n}{2}\left[\,f\big(v(t_{n+1})\big) + f\big(v(t_n)\big)\,\right] = q\big(v(t_{n+1})\big) - q\big(v(t_n)\big)$$

$$\tag{1.9}$$

where $h_n$ is the time step defined by:

$$h_n = t_{n+1} - t_n \tag{1.10}$$

Equations (1.9) can be written in the form:

$$g\,\big(\,v(t_{n+1})\,\big) = 0 \tag{1.11}$$

which is similar to equations (1.3).

The Newton-Raphson technique described in the previous section can be used to solve equations (1.11).

The right hand side vector now consists of the charges associated with each of the nodes, in addition to the independent source currents and the Norton equivalent source currents associated with each nonlinear circuit element. The Jacobian now consists of the nodal capacitance matrix, which are the derivatives of the node charges with respect to the node voltages, in addition to the the nodal conductance matrix, which are the derivatives of the node currents with respect to the node voltages. Thus at each iteration, the node currents and conductances as well as the node charges and capacitances must be recomputed for each circuit element.

### 1.3.3. Charge Nonconservation Problem

The charge nonconservation problem is briefly described here and will be addressed in more detail in a later chapter [6].

As mentioned above and indicated by equations (1.9), it is necessary to compute the charges associated with each of the terminals of the nonlinear circuit elements. Usually, it is easier to express the terminal capacitances as a function of the terminal voltages of the nonlinear elements. Thus, if only capacitance equations are available, then charges are obtained through the numerical approximation of:

$$Q = \int C(v)\, dv \qquad (1.12)$$

In this case the charges will be path dependent i.e. even if the voltage starts at a particular value and ends at the same value, the computed charge will be different if the intermediate voltage values are different. This discrepancy increases even more when the nonlinear capacitances are not controlled only by their terminal voltages alone.

This problem can be avoided by using analytical equations for $Q$ instead of numerical approximations. Thus, in order to have charge conservation, it is necessary to have charge equations instead of capacitance equations, for the nonlinear circuit elements.

### 1.3.4. AC Analysis

In the small signal ac analysis, all elements in the circuit are, by definition, linear. The nonlinear circuit elements are modeled by their equivalent linearized models, determined by the dc operating analysis. AC analysis determines the small signal solution of the circuit in sinusoidal steady state. All input sources are sinusoidal with the same input frequency, although sources may be assigned different values of relative phase.

The equations for a linear ac analysis are assembled by the same method used for the dc analysis, except for the fact that the equations are complex for ac analysis and they are linear. This results in a system of equations similar to equations (1.5) and can be solved using similar techniques.

If the circuit has only one ac input, then it is convenient to set that input to unity amplitude and zero phase. The value of each circuit variable in ac analysis then is equal to the value of the transfer function of that variable with respect to the input.

The linearized conductances and capacitances for the nonlinear circuit elements are computed at the dc operating point and the equations are assembled. Then at each frequency of interest, it is necessary to recompute only the frequency dependent conductances and capacitances.

# CHAPTER 2

# DEVICE MODELING

## 2.1. Model Development

Models describe the device behavior to the circuit simulation program. This is usually done by expressing the currents and charges associated with the device terminals in terms model equations. A device model can be used for a variety of different purposes and ideally it would be convenient to have only one model which can serve all the needs. However, different applications place different requirements on the model and many times these prove to be contradictory constraints necessitating some compromises. Different model applications and the corresponding requirements are listed below:

1.  Predicting electrical behavior of new devices:
    gain understanding of device physics

completely theoretical basis

2.    Representing devices in circuit simulation:
accuracy
simplicity to save computing time
simplicity to allow engineering understanding
minimal number of defining parameters
extending to unusable portions of device characteristics
model parameters should have direct visualizable
effect on device characteristics

3.    Process control aid:
simplicity
theoretical basis
reversible model

State of the art process technology makes it possible to
fabricate devices which encounter a variety of physical
effects. A completely theoretical model based on the fundamentals of physics becomes practically intractable. On the
other hand, use of a completely empirical model results in a
loss of predictive capabilities. A compromise is usually made
in developing models for circuit simulation. A combination
of physics based and empirical equations is used. The primary model parameters are closely linked to theory provid-
ing an engineering understanding of device physics and serv-
ing as a process control aid. The secondary model parameters are partly empirical and help keep the model equations
simple. The model equations should be formulated such
that the model parameters have direct visulizable effect on
device characteristics. It is also important to formulate the

equations such that different model parameters affect different regions of device characteristics. This results in decoupling the model parameters with minimal interaction and facilitates model parameter extraction.

The device models for circuit simulation compute the terminal currents and charges of the device as a function of the terminal voltages. These terminal currents and charges should be continuous functions of the terminal voltages for the Newton-Raphson iterations to converge. It also helps the convergence behavior if these functions have continuous first derivatives, although this is not a requirement.

The Newton-Raphson technique demands the computation of first derivatives. These derivatives are usually computed analytically. In some cases, however, the model equations are so complicated that it is tedious, cumbersome and error prone to compute analytical derivatives. Numerical derivatives using the finite difference method may be used in these situations, but this increases the model computation time.

Sometimes it is easier to divide the operating range of the device in different regions of operation so that the model equations can be conveniently formulated. If different equations are used for different regions of operation, it is important to make sure that the terminal currents and charges are continuous across the region boundaries.

During the Newton-Raphson iterations, it is possible to encounter wide variations in the terminal voltages. Therefore, it is important to consider the entire voltage range while formulating the model equations even though the device will not encounter these voltages in practical circuits.

Terminal capacitance equations are more readily available for many types of devices. However, it is necessary to formulate the charge equations if charge conservation is desired.

During model development, it is important to keep in mind the modeling paradox. More complex models can potentially represent the device characteristics more accurately. But it is more difficult to extract all the model parameters for such complex models. And therefore, if the model parameters are not specified properly, the device characteristics will not be reproduced accurately.

## 2.2. Model Specification

A program requires three types of information to specify a transistor model completely: fundamental constants, operating conditions and model parameters. The *fundamental physical constants* such as electronic charge are defined inside the circuit simulation program. The *operating conditions* define the circumstances under which the model equations are to be evaluated. In a nodal analysis program, for example, the operating conditions are normally the transistor's bias voltages. In addition to the bias voltages, the operating conditions may consist of other quantities such as the operating temperature. The third set of information required is the set of *model parameters* for each device in the circuit. A discrete circuit may contain many discrete devices of similar types. Since integrated circuit devices of similar type undergo same fabrication steps, they exhibit similar underlying behavior. Therefore, there is a large set of model parameters which are common to devices of similar type. Hence, most circuit simulations programs provide a convenient way of specifying these model parameters. Instead of specifying the same model parameters for each device, the devices can refer to a model and the parameters can then be specified for the model only once. Parameters specific to the device need only be specified with the device itself. For integrated circuit devices, these parameters are generally related to the device geometry. In this book, model parameters are denoted by uppercase words.

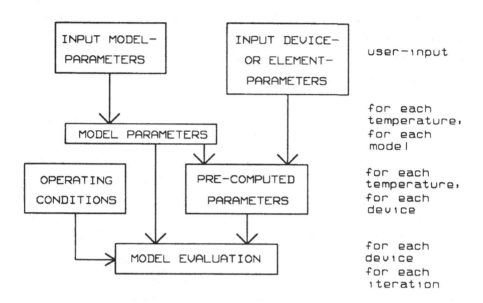

**Figure 2.1:**   Model computation flow diagram.

## 2.3. Model Computation

Various model related computations are performed in different stages of the circuit simulation program, as shown schematically in Figure 2.1. Model parameter values must be supplied by the user in a manner predetermined by the simulation program. Some programs are very flexible and provide alternate ways of specifying some model parameters. For example, the body effect factor *GAMMA* for MOSFETs depends on the substrate conecntration *NSUB*. If fabrication process information is available, then *NSUB* can be specified. If electrical measurements are available, then it is more convenient to specify *GAMMA*. The model equations use the parameter *GAMMA*, but either *GAMMA* or *NSUB* can be specified in the input. Thus a distinction must be made between the *model parameters* and the program's *input parameters*. Many model parameters depend on the operating temperature. Therefore, once the analysis temperature is known, the *model parameters* are computed based on the *input parameters* specified by the user. Some model parameters depend on the device geometry. Since these parameters are independent of the operating conditions, for computational efficiency, it is convenient to *precompute* these parameters for each device so that the same computations need not be repeated for each iteration.

Once these precomputations are done, then at each iteration, terminal currents and charges and their derivatives are computed for each device in the circuit. This is computationally most intensive part of the circuit simulation program. About 60 to 70 per cent of the analysis time is spent in model evaluation. Of course, this time depends on the circuit size and the model complexity. Therefore, it is advisable to use the simplest model that will satisfy the accuracy requirements. Many circuit simulation programs, including SPICE, use a technique to reduce the overall model evaluation time. This technique is sometimes referred to as *bypass* and involves some overhead in terms of memory usage and computation time. The operating conditions and the related current and charge information is saved for each device. If the operating conditions are not changing significantly, then the model computation is bypassed and old values are used instead, thus saving the model evaluation time. Sometimes, this may result in increasing the total number of iterations for convergence. This technique is more effective if more sophisticated tests are used to decide when the model computations can be bypassed. But this increases the model computation time and hence a compromise is necessary. Depending on the circuit activity, this technique results in 10 to 40 per cent savings in the overall model evaluation time.

## 2.4. Model Parameter Extraction

The accuracy of circuit simulation depends on the accuracy of the device models. The extraction of optimum model parameter values is necessary to ensure that the device model equations represent the device characteristics closely. Although most models are based on physical theory, there are always some parameters which do not have physically welldefined values, and others for which the physical values do not give the best fit to actual device characteristics. Thus it is generally necessary to extract model parameters from transistor data. The transistor data may be obtained from device simulation or from device characterization. Since it is more common to extract the model parameters from electrical measurements performed on a number of test devices, the transistor data will be referred as measured data.

A general purpose optimizer is extremely useful for model parameter extraction [7] - [10]. The optimizer adjusts model parameter values so that the simulated or computed device characteristics match closely with the measured data. Since most of the times the model equations are nonlinear, nonlinear optimization techniques have to be used. Although it is beyond the scope of this book to discuss optimization in detail, a brief review of the essential ideas is presented here.

In optimization problems, an objective function $F(p)$ is given, and $p^*$, the value of $p$ which minimizes $F$, is sought. If $F$ can be reasonably approximated by a quadratic near $p_k$, then it can be shown that an updated parameter vector is given by:

$$p_{k+1} = p_k - \alpha_k \ G_k^{-1} \ g_k \qquad (2.1)$$

where $g$ is the gradient of $F$ and $G$ is the matrix of second partial derivatives of $F$, referred to as the Hessian matrix. Equation (2.1) is the basis of Newton's method which is an iterative procedure for finding the minimal of $F$. The

attractive feature of Newton's method is that it exhibits quadratic convergence. A disadvantage is that it is usually
difficult and expensive to calculate the Hessian for practical
problems. A class of algorithms, known as quasi-Newton
methods, has been developed, which generates curvature
information about $F$ without having to calculate second
derivatives.

For many practical problems, including parameter
extraction, the natural choice for the objective function is a
least squares form:

$$F(p) = \frac{1}{2}\, r^T(p)\, r(p) \qquad\qquad (2.2)$$

where $r$ is the vector of residuals. For the least squares
case,

$$g(p) = J^T(p)\, r(p) \qquad\qquad (2.3)$$

and,

$$G(p) = J^T(p)\, J(p) + Q(p) \qquad\qquad (2.4)$$

where $J$ is the Jacobian matrix whose columns are the gra-
dients of the components of $r$ with respect to $p$, and $Q$ is a
matrix formed by calculating the product of each component
of $r$ with the Hessian matrix of that component and then
summing these matrices together. If the residuals are small,
then $Q$ can be neglected. This is the Gauss-Newton method
which has the interesting property that the Hessian involves
only first derivatives. A generalization of the Gauss-Newton
method is the popular Levenberg-Marquardt method [11]
which adds a constant diagonal matrix to the first order
approximation to the Hessian. The constant is called the
Marquardt parameter. When the Marquardt parameter is
small, this reduces to the Guass-Newton method with its
rapid convergence, and when the Marquardt parameter is
large, the method becomes a steepest descent algorithm,
with its inherent stability. Both these methods assume that

eventually $J^T J$ will be a good approximation to the Hessian. For large residual problems where $F(p_{k+1})$ is not small, this is not true and it is possible to augment the Hessian with quasi-Newton updates.

The derivatives of model equations with respect to the model parameters need to be computed for this method. Many times these derivatives are not readily available. The most obvious alternative is to use finite difference approximations. Careful attention must be paid to the choice of interval and the method (forward, central or backward differences) by which the derivatives are computed. This is particularly true when the iteration sequence comes close to the optimal value, and the true value of the gradient is small. In these cases sometimes it is advantageous to switch from an easily computed forward difference method to the more expensive but more accurate central difference approximation.

For the model parameter extraction problem, the least squares objective function can be written in a general form:

$$F(p) = \sum_n \left\{ \left[ I_{meas}(n) - I(p,n) \right] w(n) \right\}^2 \qquad (2.5)$$

where $p$ represents the parameter vector which characterizes the transistor model, $n$ represents the entire set of simulation or measurement conditions, i.e., range of transistor geometries and terminal voltage conditions, $I_{meas}(n)$ is the measured or simulated data at $n$ and $I(p,n)$ are the computed values from the transistor model at $p$ and $n$. The weighting function $w(n)$ can be used to weight critical regions of device performance so that the model is forced to fit adequately the data in these regions. The simplest weighting function can be unity in which case the weighting is performed implicitly by the distribution of data points. A more commonly used weighting function is based on some minimum data value $I_{\min}$ such that,

$$w(n) = \frac{1}{\max\left(I_{meas}(n), I_{\min}\right)} \qquad (2.6)$$

Figure 2.2 illustrates the flow chart of a typical parameter extraction program. The program begins with the

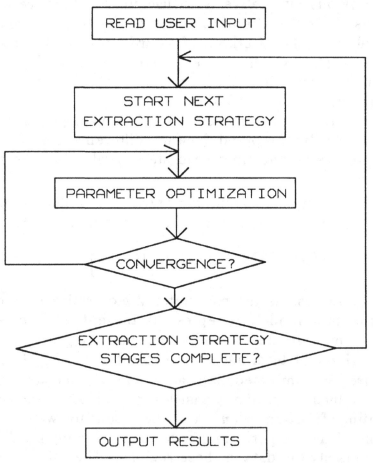

**Figure 2.2:**    Flow chart of a typical parameter extraction program.

initialization of the user input parameters. After initialization, the first extraction stage is begun. An optimal search direction and step size are chosen by the optimization algorithm and an error function and gradient are evaluated in a multidimensional function space. The next program stage is to check for convergence. If the error function residuals meet the convergence criteria, then the program continues on to the next extraction stage. If not, then the program iterates until the convergence criteria are met or until the maximum allowed number of iterations has been reached. The extraction procedure continues until all of the extraction stages have been completed. The user inputs provide a mechanism for controlling the parameter extraction program at execution time. The user inputs may be supplied either as responses to a menu driven user interface in the manual mode, or from user input files that have been prepared beforehand in the automatic mode. During each extraction stage, the user may specify which device parameters are held constant and which are allowed to vary. Several different parameter optimization and error minimization algorithms may be available to the user. The extraction strategy should allow the user to select among a manual mode, a semi automatic mode and a fully automated mode of operation. Typically, the manual mode of operation is useful during the initial work of developing a new device model. The automatic mode of operation is advantageous later after it has become clear which parameter extraction procedure will produce the best results for given devices and technology.

To be useful as a production tool, any extraction system must produce parameters reliably and accurately. Two potential sources of error in the optimization technique described here are:

1.    Local minima,
2.    Redundancy of parameters.

The first can result in nonoptimum parameter values; the second can produce nonunique parameter values, but gives an optimum fit to the data.

The optimization algorithm always converges towards a minimum total error with the stability of the steepest descent method. The accuracy in locating the minimum is primarily determined by the convergence criteria - tighter tolerances result in more iterations and less error. The maximum obtainable accuracy is limited by errors in the numerical calculation of the gradient, and hence of the Jacobian. The relative error in the gradient becomes large as its value approaches zero. In practice it has been found that the uncertainty in the extracted parameters due to errors in locating the minimum is small compared to the uncertainty arising from other causes such as the choice and weighting of data points. The extraction procedure has no provision for distinguishing between local and global minima. Thus it is possible that a local minimum may be found, if one exists.

The simplest case of redundant parameters occurs when the function to be fit is independent of the value of the parameter being extracted. In this case, any value of the parameter gives the same characteristics. The total error is minimized, but the parameter value is not unique. If the function is only weakly dependent on a parameter, small errors in locating the minimum can produce large errors in the value of the parameter. Again, although the extracted parameter values are not unique, the fit of the data is still optimum. The nonuniqueness of parameters may also be produced by local minima; the two problems thus may often be confused.

The hazards imposed by local minima and redundant parameters are illustrated by the following equation, which is the mobility reduction equation for the SPICE level 2 MOSFET model.

$$\mu_{eff} = \begin{cases} UO & \text{if } V_{GS} - V_T \leq V_{CRIT} \\ UO \left[ \dfrac{V_{CRIT}}{V_{GS} - V_T} \right]^{UEXP} & \text{if } V_{GS} - V_T > V_{CRIT} \end{cases}$$

$$(2.7)$$

where,

$$V_{CRIT} = \frac{\epsilon_{si}}{C_{OX}} \, UCRIT \qquad (2.8)$$

$UCRIT$, $UEXP$ and $UO$ are the parameters to be extracted. For $V_{GS} - V_T > V_{CRIT}$, the parameters $UCRIT$ and $UO$ are redundant, in that for any value of $UCRIT$, $UO$ may be adjusted to give the same minimum error. It also happens that for some sets of data, increasing $UCRIT$ to the point that $V_{GS} - V_T$ becomes less than $V_{CRIT}$ in fact reduces the error. Thus the "plateau" in the error versus $UCRIT$ function is a special case of a local minimum. Three solutions to the problem of the plateau are available: 1) provide data at values of $V_{GS}$ low enough so that $V_{GS} - V_T < V_{CRIT}$ for reasonable values of $UCRIT$; 2) specify an initial value of $UCRIT$ such that $V_{GS} - V_T < V_{CRIT}$ and/or 3) use a better behaved equation for mobility reduction.

Thus, avoiding the hazards of parameter extraction requires careful choice of:

1.   measured data,

2.   parameters to be extracted,

3.   initial values of parameters.

Although the general purpose model parameter extraction program is a very powerful and useful tool, it must be used with extreme care.

A well formulated model contains majority of model parameters which affect device characteristics in specific

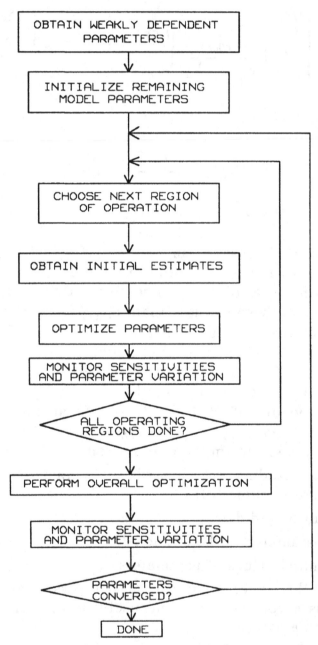

**Figure 2.3:**    Model parameter extraction procedure.

regions of operation or which dominate behavior of devices having specific range of geometries. There are only a few parameters which affect more than one regions of operation. Therefore, it is recommended that parameters should be extracted from the region of operation in which those parameters dominate the device behavior. Good initial estimates can be obtained by simplifying the model equations for the region of operation of interest such that closed form solution may be obtained for the model parameter values in terms of measured data and other known model parameters. Model parameters are extracted sequentially, in small groups. The value of each parameter is fixed for use in extracting further parameters. A small portion of the model equations and data from a limited part of a device's operating range are used in extracting each group of parameters. Once this sequential, regionalized extraction is completed, the model parameters which affect more than one region of operation are extracted. Measured data covering all regions of operation is then used and all the model parameters are allowed to vary to obtain the best overall fit. This accounts for the model parameter interaction as well as for the model parameters which affect the device characteristics in regions of operation other than the ones from which they were extracted earlier. This entire procedure is then repeated until consistent values are obtained for all model parameters. At each extraction stage, the sensitivity of the error function to the extracted model parameters is monitored to avoid the hazards mentioned earlier. The difference between the final optimized parameter values and their corresponding initial estimates should also be monitored. Small sensitivites or large differences should lead to the re-examination of the overall extraction strategy in terms of regrouping the parameters or changing the choice of the measured data. The model equations are very weak functions of some of the model parameters. The model parameter $PHI$ for the

MOSFET model is an example of this type of parameter. Sometimes it is recommended that the values of these parameters be obtained by some other means. For example, the model parameter *PHI* can be determined from the fabrication process information. The model parameter extraction procedure is shown schematically in Figure 2.3.

# CHAPTER 3

# DIODE MODELS

## 3.1. Introduction

Although diode is not a field effect device, it is part of the parasitics associated with the field effect devices. Field effect devices have junction diodes which are not part of the intrinsic device but are part of the parasitic components which need to be included in the model. This chapter describes the diode model suitable fo this purpose. This model is simpler than the model implemented as a general diode model in many of the circuit simulation programs. The general diode model includes parasitic series resistance effects and reverse bias breakdown effects which are not modeled for the parasitic diodes. For normal device operation, the parasitic diodes are supposed to be reverse biased and their contribution to the device characteristics is normally negligible. Hence there is no need to use a complicated model. Following physical effects are not included in the simple diode model:

1.    Parasitic series resistance
2.    Reverse bias breakdown
3.    High level injection
4.    Junction recombination at low forward bias.

Most of the general diode models include the first two effects but do not model the last two effects.

SPICE uses different model equations for the diode current for MOSFETs, JFETs and MESFETs [3]. However, to make the simulation program modular and easy to maintain, it is advisable to write only one function to perform the parasitic diode computations. This function can then be called by all the other models which need to include the parasitic diodes.

The diode equivalent circuit is shown schematically in Figure 3.1.

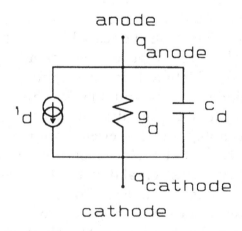

**Figure 3.1:**    Diode equivalent circuit.

## 3.2. DC Characteristics

The dc model equations express the dc current through the diode as a function of the voltage across the diode. The diode conductance, which is the derivative of the diode current with respect to the diode voltage, is also computed.

To aid convergence, SPICE adds a small conductance in parallel with every junction diode. The value of this conductance, *GMIN*, is a program parameter that can be set by the user. The default value for *GMIN* is $10^{-12}$ S. Some circuit simulation programs do not use this conductance.

### 3.2.1. MOSFET diodes

Following equations are used for the parasitic diodes associated with the MOSFETs in SPICE:

For $v_d > 0$,

$$i_d = IS \left( e^{\frac{v_d}{v_t}} - 1 \right) \tag{3.1}$$

$$g_d = \frac{IS}{v_t} e^{\frac{v_d}{v_t}} + GMIN \tag{3.2}$$

For $v_d \leq 0$,

$$i_d = \frac{IS}{v_t} v_d \tag{3.3}$$

$$g_d = \frac{IS}{v_t} + GMIN \tag{3.4}$$

where,

   $i_d$ =current through the diode

   $v_d$ =voltage across the diode

   $g_d$ =diode conductance

$$v_t = \frac{k\,T}{q}$$

where $k$ is the Boltzmann's constant, $q$ is the electronic charge and $T$ is the diode temperature in degrees Kelvin. *IS* is the saturation current and is a user input model parameter. Figure 3.2 shows these characteristics.

In the reverse bias region, the diode is modeled as a constant conductance of value $IS\,/\,v_t$. This over estimates the current through the diode in the reverse direction and this model does not agree with the measured diode characteristics. The finite conductance in the reverse direction aids the convergence of the simulation algorithm, as opposed to the measured conductance of zero value, which may cause convergence problems for some, but not necessarily all,

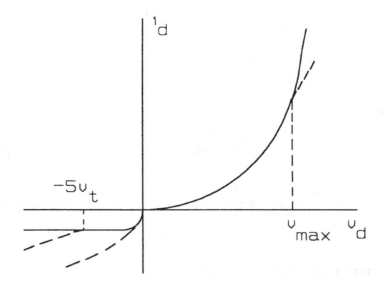

**Figure 3.2:**    Measured and modeled diode characteristics.

simulation algorithms. The current and conductance are continuous at $v_d = 0$, but there is an inconsistency in the $i_d$ and $g_d$ equations so that:

$$g_d \neq \frac{\partial i_d}{\partial v_d}$$

### 3.2.2.  JFET and MESFET diodes

SPICE uses the following equations for the parasitic diodes:

For $v_d \geq - 5 \ v_t$,

$$i_d = IS \left( e^{\frac{v_d}{v_t}} - 1 \right) + GMIN \ v_d \qquad (3.5)$$

$$g_d = \frac{IS}{v_t} \ e^{\frac{v_d}{v_t}} + GMIN \qquad (3.6)$$

For $v_d < - 5 \ v_t$,

$$i_d = - \ IS + GMIN \ v_d \qquad (3.7)$$

$$g_d = - \ \frac{IS}{v_d} + GMIN \qquad (3.8)$$

where,

$i_d$ = current through the diode

$v_d$ = voltage across the diode

$g_d$ = diode conductance

$$v_t = \frac{k \ T}{q}$$

where $k$ is the Boltzmann's constant, $q$ is the electronic charge and $T$ is the diode temperature in degrees Kelvin. $IS$ is the saturation current and is a user input model parameter.

Figure 3.2 shows these characteristics.

In the reverse bias region, the model equations are not consistent, so that:

$$g_d \neq \frac{\partial i_d}{\partial v_d}$$

Also the reverse current is over estimated compared to the measured data. The current and conductance equations are discontinuous at $v_d = -5 \; v_t$.

### 3.2.3. Parasitic diode model

A recommended parasitic diode model is given below:

For $v_d > v_{\max}$,

$$i_d = IS \; e^{\frac{v_{\max}}{v_t}} \left[ 1 + \frac{v_d - v_{\max}}{v_t} + e^{-\frac{v_{\max}}{v_t}} \right] \qquad (3.9)$$

$$g_d = \frac{IS}{v_t} \; e^{\frac{v_{\max}}{v_t}} \qquad (3.10)$$

For $-5 \; v_t < v_d \leq v_{\max}$,

$$i_d = IS \left( e^{\frac{v_d}{v_t}} - 1 \right) \qquad (3.11)$$

$$g_d = \frac{IS}{v_t} \; e^{\frac{v_d}{v_t}} \qquad (3.12)$$

For $v_d \leq -5 \; v_t$,

$$i_d = IS \; e^{-5} \left[ 1 - e^5 + 5 + \frac{v_d}{v_t} \right] \qquad (3.13)$$

$$g_d = \frac{IS}{v_t} \; e^{-5} \qquad (3.14)$$

where,

$i_d$ = current through the diode

$v_d$ = voltage across the diode

$g_d$ = diode conductance

$v_t = \dfrac{k\ T}{q}$

$v_{\max} = v_t\ MAX\_EXP$

where $k$ is the Boltzmann's constant, $q$ is the electronic charge and $T$ is the diode temperature in degrees Kelvin. *IS* is the saturation current and is a user input model parameter. Figure 3.2 shows these characteristics.

This model is continuous in both $i_d$ and $g_d$ across all the regions of bias voltages. This model suffers from the same drawback of previous models and over estimates the current for $v_d \leq -5\ v_t$. If this is not desirable, then the exponential equation used for $v_d \leq v_{\max}$ can be used in the reverse bias region also. Also, a conductance of *GMIN* may be added to $g_d$ and the corresponding *GMIN* $v_d$ contribution may be added to $i_d$ if it is required by the simulation algorithm. In the high forward bias region, this model uses a linear approximation to $i_d$. This is done to limit the exponentially dependent current so as not to exceed the computer's range of floating point number representations. *MAX_EXP* is a constant such that $e^{MAX\_EXP}$ is less than but close to the maximum floating point number. Usually $v_{\max}$ is much higher than any voltage encountered in practical circuits and hence the linear approximation does not introduce any errors in representing the measured diode characteristics but avoids potential numerical errors in the Newton-Raphson algorithm. If more flexibility is desired in representing the forward bias characteristics, then,

$$v_t = \frac{N \, k \, T}{q} \qquad (3.15)$$

may be used, where $N$ is the diode ideality factor which determines the slope of the $\ln \left( i_d \right)$ versus $v_d$ characteristics in the forward bias region.

### 3.2.4. Schottky barrier diodes

SPICE uses the same model for p-n junction diodes and Schottky barrier diodes. However, Schottky diodes associated with GaAs MESFETs have different behavior compared to the p-n junction diodes [12]. The reverse diode current increases rapidly for low reverse bias and then increases monotonically with a smaller slope for large reverse bias. This is contrary to the behavior found in a classical diode, where the reverse current remains constant until a breakdown occurs. The slower increase is caused by the lowering of the Schottky barrier under reverse bias conditions. An explanation for the rapid initial reverse current increase is not readily available, but a change in the dominant conduction mechanism is suggested as a possible reason. The relatively small Schottky barrier height and the comparatively large low field electron mobility in GaAs suggest that this may be caused by electronic tunneling. An accurate model for the parasitic diodes becomes important for enhancement type GaAs MESFETs where the diode current contribution may be significant.

The following equations are suggested for the Schottky barrier diodes [12]:

For $v_d \geq 0$,

$$i_d = ISF \left( e^{\frac{v_d}{N \, v_t}} - 1 \right) \qquad (3.16)$$

For $v_d < 0$,

$$i_d = ISR \ v_d \ e^{-\frac{DELTA \ v_d}{v_t}} \qquad (3.17)$$

where,

$i_d$ = current through the diode

$v_d$ = voltage across the diode

$g_d$ = diode conductance = $\dfrac{\partial i_d}{\partial v_d}$

$v_t = \dfrac{k \ T}{q}$

where $k$ is the Boltzmann's constant, $q$ is the electronic charge and $T$ is the diode temperature in degrees Kelvin. $ISF$ and $ISR$ are model parameters representing the saturation current in the forward and reverse directions respectively. $N$ represents the diode ideality factor and $DELTA$ is the reverse bias Schottky barrier lowering coefficient. Although $i_d$ is continuous at $v_d = 0$, there is a discontinuity in $g_d$.

Another approach is to model the Schottky diode using the equivalent circuit shown in Figure 3.3 [13]. This model accounts for the nonlinear series resistance associated with the Schottky diode and can predict the forward bias characteristics accurately. It introduces one additional nonlinear element and one additional node for each diode.

## 3.3. Junction voltage limiting

The exponential nonlinearities associated with the diode models can result in slow convergence of the Newton-Raphson iterations or the iteration procedure may even be stopped when numbers overflow computer's arithmetic capabilities [14]. Several techniques have been used to limit the

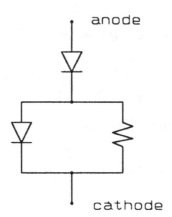

**Figure 3.3:**    Schottky diode model.

excursion of junction voltage such that its exponentially dependent current does not exceed the computer's range of floating point number representations. These techniques fall into two general categories: node voltage limiting and branch voltage limiting.

Node voltage limiting techniques simply limit the change in each node voltage from iteration to iteration to some predetermined small value. Branch voltage limiting techniques limit the voltage change across individual branches in the circuit. Branch characteristics can be taken into consideration in designing the limiting schemes for different types of branches. Several branch voltage limiting techniques [14] are described in this section.

The problem of branch voltage limiting is illustrated in Figure 3.4. It is assumed that the circuit has been solved at the point $( v_o , i_o )$ and that the solution of the resulting linearized circuit yields $\hat{v}$ for use in the next iteration. It is desired to choose a new point for linearization $v_1 \leq \hat{v}$ such that $i_1$ is representable within the range of valid floating point numbers.

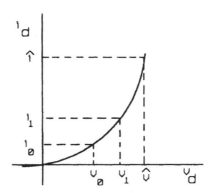

**Figure 3.4:**   Junction voltage limiting for diodes.

The simplest technique is to place fixed bounds on the excursions $\hat{v} - v_o$. One of the most effective forms of the fixed bound approach is summarized below:

$$
v_1 = \begin{cases}
\hat{v} & \text{if } \left| \hat{v} - v_o \right| \leq 2\,v_t \\
\hat{v} & \text{if } \hat{v} \leq 10\,v_t \text{ and } v_o \leq 10\,v_t \\
v_o - 2v_t & \text{if } \hat{v} < v_o \text{ and } 10\,v_t \leq v_o \\
\max\left( 10\,v_t \,,\, v_o + 2\,v_t \right) & \text{if } v_o < \hat{v} \text{ and } 10\,v_t < \hat{v}
\end{cases}
$$

$$(3.18)$$

In essence, for voltages less than $10\,v_t$, no limiting is used, (i.e. $v_1 = \hat{v}$) while for voltages greater than $10\,v_t$, excursions are limited to $v_o \pm 2\,v_t$.

A second fixed bound approach makes use of the hyperbolic tangent function and is summarized below:

$$
v_1 = v_o + 10\,v_t \, \tanh\left( \frac{\hat{v} - v_o}{10\,v_t} \right) \qquad (3.19)
$$

Since the hyperbolic tangent function ranges from -1 to +1, the maximum excursion of $v_1$ is $v_o \pm 10\, v_t$. For small excursions, $v_1 \approx \hat{v}$ since the slope of the hyperbolic tangent function for small arguments is one.

The alternating bias technique uses iteration on current for increasing voltages and iteration on voltage for decreasing voltages:

$$v_1 = \begin{cases} \hat{v} & \text{if } \hat{v} \leq 0 \text{ or } \hat{v} \leq v_o \\ v_o + v_t \ln \left( \dfrac{\hat{v} - v_o}{v_t} + 1 \right) & \text{if } \hat{v} > 0 \text{ or } \hat{v} > v_o \end{cases}$$

$$(3.20)$$

A modification of the above method is used in SPICE. A current iteration is used whenever the diode current of the new iterate has a slope larger than a specified value, and a voltage iteration is used otherwise. Since the slope can be uniquely related to a voltage, it is sufficient to define a critical voltage $v_{crit}$ rather than compute the slope at each iteration. $v_{crit}$ is chosen as the voltage at which the diode equation has a minimum radius of curvature.

$$v_{crit} = v_t \ln \left( \frac{v_t}{\sqrt{2}\, IS} \right) \qquad (3.21)$$

$$v_1 = \begin{cases} \hat{v} & \text{if } \hat{v} \leq v_{crit} \\ v_o + v_t \ln \left( \dfrac{\hat{v} - v_o}{v_t} + 1 \right) & \text{if } \hat{v} > v_{crit} \quad (3.22) \end{cases}$$

In SPICE implementation, $v_{crit}$ is computed based on the model parameter $IS$, but the effective $IS$ for the diode is obtained by multiplying the model parameter $IS$ by the diode area factor. Therefore, a proper implementation would involve computing $v_{crit}$ for each diode in the circuit as opposed to the present computation of $v_{crit}$ done for each diode model.

### 3.4. Charge storage

Two distinct mechanisms contribute to the diode charge storage. Charge storage due to injected minority carriers depends on the current flowing through the diode and is given by the diffusion capacitance equations:

$$q_{anode} = TT \; i_d \qquad (3.23)$$

$$c_d = TT \; g_d \qquad (3.24)$$

$$q_{cathode} = - \; q_{anode} \qquad (3.25)$$

where,

$q_{anode}$ = charge associated with the anode

$q_{cathode}$ = charge associated with the cathode

$c_d$ = diode capacitance = $\dfrac{\partial q_{anode}}{\partial v_d}$

$TT$ is the model parameter representing the transit time across the diode. Since the parasitic diodes are normally reverse biased, the diode current is usually small and hence this charge contribution is normally not included in the parasitic diode model.

Charge storage in the junction depletion region is represented by a capacitive component:

$$c_d = \frac{CJO}{\left(1 - \dfrac{v}{PB}\right)^M} \quad \text{if } v < FC\ PB$$

(3.26)

$$c_d = \frac{CJO}{\left(1 - FC\right)^{(1 + M)}} \left[1 - FC\left(1 + M\right) + \frac{v}{PB}\ M\right]$$

$$\text{if } v \geq FC\ PB \quad (3.27)$$

$$q_{anode} = \int_0^v C_d(v)\ dv \quad\quad\quad (3.28)$$

$$q_{cathode} = -\ q_{anode} \quad\quad\quad (3.29)$$

where,

$q_{anode}$ = charge associated with the anode

$q_{cathode}$ = charge associated with the cathode

$c_d$ = diode capacitance

$CJO$ is the junction depletion capacitance when $v_d = 0$, $PB$ is the built-in potential across the junction and $M$ is the grading coefficient of the junction. For $v_d > 0$, equation (3.25) becomes numerically unstable and in fact becomes singular at $v_d = PB$. This singularity is avoided by approximating the capacitance by a straight line for $v_d > FC\ PB$, where $FC$ is a user selectable model parameter, which is usually set to 0.5. The junction depletion capacitance actually decreases at high forward bias, as shown in Figure 3.5. The model deviates from this behavior but it does not affect the accuracy of simulation since the junction depletion capacitance does not affect the overall diode behavior in the forward bias region. Since appreciable current flows through the diode when it is forward biased, diode behavior is dominated by the conductance and the diffusion capacitance.

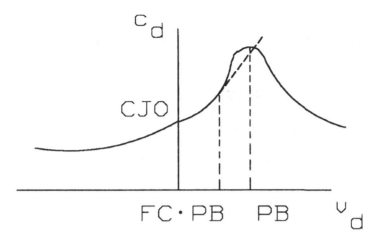

**Figure 3.5:**    Junction depletion capacitance.

### 3.5.  Temperature dependence

The model parameters $IS$, $PB$ and $CJO$ are functions of the diode temperature.

$$IS \; \alpha \; e^{-\frac{E_g}{kT}} \qquad (3.30)$$

$$IS = IS_{TNOM} \, e^{-\left(\frac{E_g}{kT} - \frac{E_{g\,TNOM}}{kT}\right)} \qquad (3.31)$$

where,

$IS_{TNOM} = IS$ at nominal temperature $TNOM$

$IS = IS$ at temperature $T$

$E_{g\,TNOM} =$ band gap at temperature $TNOM$

$E_g =$ band gap at temperature $T$

$TNOM =$ nominal temperature in degrees Kelvin

$T =$ analysis temperature in degrees Kelvin

The band gap is given by:

$$E_g = 1.16 - \frac{7.02 \times 10^{-4} \ T^2}{T + 1108.0} \quad (eV) \quad \text{for} \quad Si \quad (3.32)$$

$$E_g = 1.52 - \frac{5.41 \times 10^{-4} \ T^2}{T + 204.00} \quad (eV) \quad \text{for} \quad GaAs \quad (3.33)$$

The built-in potential $PB$ is given by:

$$PB = \frac{kT}{q} \ln \frac{N_A \ N_D}{n_i^{\ 2}} \quad (3.34)$$

where,

$$n_i \ \alpha \ T^{1.5} \ e^{-\frac{E_g}{2kT}} \quad (3.35)$$

Therefore, $PB$ can be written as:

$$PB = 2 \ \frac{T}{TNOM} \ PB_{TNOM}$$

$$- 2 \ \frac{kT}{q} \ \left[ \frac{3}{2} \ \ln \frac{T}{TNOM} + \frac{1}{2k} \left( - \frac{E_g}{T} + \frac{E_{g \ NOM}}{TNOM} \right) \right]$$

$$(3.36)$$

SPICE uses the following empirical equation for the junction capacitance:

$$CJO = CJO_{NOM}$$

$$\left\{ 1 + M \left[ 0.0004 \ (T - TNOM) - \frac{PB - PB_{TNOM}}{PB_{NOM}} \right] \right\}$$

$$(3.37)$$

An alternate equation is:

$$depletion \ region \ width \ \alpha \ PB^M$$

$$CJO \ \alpha \ \frac{1}{depletion \ region \ width}$$

$$CJO \; \alpha \; \frac{1}{PB^M} \tag{3.38}$$

$$CJO = CJO_{TNOM} \left[ \frac{PB_{TNOM}}{PB} \right]^M \tag{3.39}$$

### 3.6. Model parameter extraction

Since the parasitic diodes associated with the active field effect devices have very small geometries, it is difficult to perform accurate measurements of the diode characteristics. Therefore, usually, special diode structures are fabricated on a test chip for determining the model parameters.

The model parameters for the dc model are extracted from the dc measurements of the forward characteristics. Since the diode current increases exponentially with the diode voltage, it is preferable to measure the dc characteristics by forcing a current through the diode and measuring the voltage across the diode. In the forward bias region, the diode current can be expressed as:

$$\ln \left( i_d \right) = \ln \left( IS \right) + \frac{v_d}{N \, v_t} \tag{3.40}$$

and is shown graphically in Figure 3.6. It is important to make measurements in the linear region of the characteristics. The characteristics deviate from the linear behavior at very low forward bias due to leakage current and junction recombination current contributions. At high forward bias, the characteristics deviate from the linear behavior due to high level injection effects and due to the presence of parasitic series resistance. None of these effects is included in the model. $IS$ is determined from the intercept of the straight line with the $\ln \left( i_d \right)$ axis and the ideality factor $N$ is determined from the slope of the straight line. $N$ is usually close to unity.

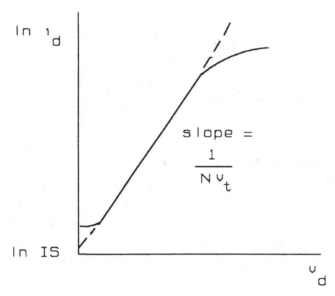

**Figure 3.6:**    A semi logarithmic plot of forward bias dc
diode characteristics.

The junction capacitance parameters are determined by
measuring the junction capacitance over a range of reverse
bias voltages and fitting the following equation to the measured data:

$$c_d = \frac{CJO}{\left(1 - \dfrac{v}{PB}\right)^M} + c_{stray} \qquad (3.41)$$

where $c_{stray}$ is stray capacitance which represents the component of $c_d$ which is not a function of the voltage across the
diode. It is important to make sure that the stray capacitance
is negligible over the entire measurement range, or it is
estimated accurately since other model parameters are sensitive to the value of the stray capacitance used in equation
(3.41). Actually, $c_{stray}$ is a redundant parameter in equation
(3.41). Therefore, it is recommended that the stray capacitance be obtained by other methods and not determined by

curve fitting the above equation with $c_{stray}$ treated as one more parameter to be extracted. *CJO* can be computed directly from the measured capacitance with zero applied bias. Parameters $M$ and *PB* are determined by using non-linear optimization. $M$ is 1/3 for a linearly graded junction and 1/2 for an abrupt step junction, but it is not uncommon to obtain values outside this range.

# CHAPTER 4

# JFET MODELS

## 4.1. Introduction

The junction field effect transistor (JFET) is basically a voltage controlled resistor [15]. Because its conduction process involves predominantly one kind of carrier, the JFET is called a ''unipolar'' transistor to distinguish it from the bipolar junction transistor (BJT), in which both types of carriers are involved.

Field effect transistors offer many attractive features for applications in analog switching, high input impedance amplifiers, microwave amplifications and integrated circuits. The FETs have considerably higher input impedance than bipolar transistors, which allows the input of a FET to be more readily matched to the desirable system impedance. The FET has a negative temperature coefficient at high current levels; that is, the current decreases as temperature increases. This characteristic leads to a more uniform temperature distribution over the device area and prevents the FET from thermal runaway or second breakdown, which occurs in the bipolar transistor. The device is thermally stable, even when active area is large or when many devices are connected in parallel. Because FETs are unipolar devices, they do not suffer from minority carrier charge storage effects, and consequently have higher switching speeds and higher cutoff frequencies. Since the devices are basically square law or linear devices, intermodulation and cross modulation distortion products are much smaller than those of bipolar transistors.

## 4.2.  DC Characteristics

The JFET consists of a conductive channel made of n-type material, with two ohmic contacts, one acting as the source and the other as the drain. When a positive voltage is applied to the drain with respect to the source, electrons flow from source to drain. The third electrode, the gate, forms a rectifying junction with the channel. The device is basically a voltage controlled resistor, and its resistance can be varied by varying the width of the depletion layer extending into the channel region.

The basic current-voltage characteristics of a JFET are shown in Figure 4.1. The characteristics can be divided into three regions. In the linear region, the drain voltage is small and $i_{ds}$ is proportional to $v_{ds}$. The current remains essentially constant and independent of $v_{ds}$ in the saturation region. In the breakdown region, the drain current increases rapidly with a slight increase of the drain to source voltage. As the reverse gate bias increases, both the saturation current $i_{dsat}$ and the corresponding saturation voltage $v_{dsat}$ decrease due to the reduced initial channel thickness which results in larger initial channel resistance. It is observed that

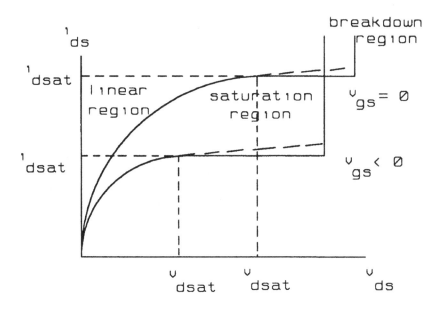

**Figure 4.1:** Current-voltage characteristics of a JFET.

for actual devices, the drain current increases slightly with increasing drain to source voltage in the saturation region as shown by the dotted lines in Figure 4.1. Since the devices are not operated in the breakdown region in practical circuits, this region of operation is not included in the JFET model.

The JFET equivalent circuit is shown in Figure 4.2. $RS$ and $RD$ are the parasitic resistances and the p-n junctions between the gate and the source and drain terminals are represented by the two parasitic diodes. The drain current is given by,

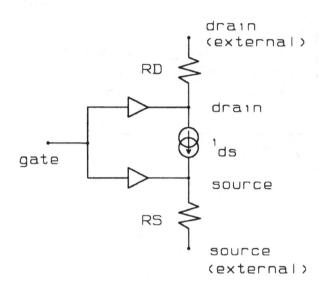

**Figure 4.2:**    JFET equivalent circuit.

If $v_{gs} - VTO \leq 0$,

    $i_{ds} = 0$

If $0 < v_{gs} - VTO < v_{ds}$,

    $i_{ds} = BETA \left( v_{gs} - VTO \right)^2 \left( 1 + LAMBDA\ v_{ds} \right)$

If $0 < v_{ds} \leq v_{gs} - VTO$,

$i_{ds} = 2\ BETA \left( 1 + LAMBDA\ v_{ds} \right) v_{ds} \left( v_{gs} - VTO - \dfrac{1}{2}\ v_{ds} \right)$

$$(4.1)$$

where,

$$v_{dsat} = v_{gs} - VTO$$

The equations are derived under the following assumptions: (1) gradual channel approximation, (2) abrupt depletion layer, and (3) constant mobility. The convention in SPICE is that VTO is negative for all JFETs regardless of polarity. In practice, however, VTO is negative for n-channel devices and positive for p-channel devices.

## 4.3. Device symmetry

Equations (4.1) are valid only for $v_{ds} \geq 0$. For $v_{ds} < 0$, the role of the source and the drain terminals is reversed. Thus, among the source and drain terminals, the terminal at a lower potential is treated as the source terminal. This assumes that the device is symmetrical with respect to the source and the drain terminals. Care must be taken when computing $i_{ds}$ for $v_{ds} < 0$. If $v_{ds} < 0$, then $v_{ds}$ is replaced by $-v_{ds}$ and $v_{gs}$ is replaced by $v_{gd}$ in equations (4.1). After the model computation, the sign of $i_{ds}$ and its derivatives is reversed. Care must also be taken in loading the conductances in the proper locations of the Jacobian or the Y matrix. Some versions of the SPICE programs do not load the conductances in their proper locations when $v_{ds} < 0$, which can cause convergence problems.

## 4.4. Voltage limiting

The concept of voltage limiting was explained earlier. SPICE uses branch voltage limiting for gate-source and gate-drain voltages. These equations are completely empirical and are based on heuristics. Some other simulators do not use limiting or use much simpler limiting equations without encountering any numerical problems.

### 4.4.1. Gate-source voltage

$v_{gso}$ is the old value of $v_{gs}$. The solution of the circuit yields $\hat{v}_{gs}$ for use in the next iteration. It is desired to choose $v_{gs1} \leq \hat{v}_{gs}$. Following intermediate quantities are used:

$$\Delta v_{gs} = \left| \hat{v}_{gs} - v_{gso} \right|$$

$$VTO_x = VTO + 3.5$$

$$v_{hi} = 2 \left| v_{gso} - VTO \right| + 2$$

$$v_{lo} = \left| v_{gso} - VTO \right| + 3$$

where $VTO$ is the threshold voltage.

If $v_{gso} < VTO$ ( *off* region) and $\hat{v}_{gs} > v_{gso}$ ($v_{gs}$ increasing):

$$v_{gs1} = \begin{cases} VTO + 0.5 & \text{if } \hat{v}_{gs} > VTO + 0.5 \\ v_{gso} + v_{lo} & \text{if } \Delta v_{gs} > v_{lo} \\ \hat{v}_{gs} & \text{if } \Delta v_{gs} \leq v_{lo} \end{cases}$$

If $v_{gso} < VTO$ ( *off* region) and $\hat{v}_{gs} \leq v_{gso}$ ($v_{gs}$ decreasing):

$$v_{gs1} = \begin{cases} v_{gso} - v_{hi} & \Delta v_{gs} > v_{hi} \\ \hat{v}_{gs} & \Delta v_{gs} \leq v_{hi} \end{cases}$$

If $v_{gso} < VTO_x$ ( *intermediate* region):

$$v_{gs1} = \begin{cases} \min \left( \hat{v}_{gs} , VTO + 4 \right) & \text{if } \hat{v}_{gs} > v_{gso} \\ \max \left( \hat{v}_{gs} , VTO - 0.5 \right) & \text{if } \hat{v}_{gs} \leq v_{gso} \end{cases}$$

If $v_{gso} \geq VTO_x$ ( *on* region):

$$
v_{gs1} = \begin{cases}
v_{gso} + v_{hi} & \text{if } \hat{v}_{gs} > v_{gso} \\
\max \left( \hat{v}_{gs} , VTO + 2 \right) & \text{if } \hat{v}_{gs} < VTO_x \\
v_{gso} - v_{lo} & \text{if } \hat{v}_{gs} > v_{lo} \\
\hat{v}_{gs} & \text{if } \hat{v}_{gs} \leq v_{lo}
\end{cases}
$$

### 4.4.2. Gate-drain voltage

The equations for gate-drain voltage are identical to the equations used for gate-source voltage, except of course gate-source voltage is replaced by gate-drain voltage.

### 4.5. Charge storage

Charge storage in a JFET occurs in the two parasitic diodes associated with the gate-source and gate-drain junctions and it is modeled using the diode model described earlier.

### 4.6. Temperature dependence

SPICE only accounts for the temperature dependence of the parameters related to the parasitic diodes. Since $BETA$ is related to the mobility of the carriers, which is a function of temperature, actually $BETA$ should also be treated as a function of temperature.

$$
mobility \; \alpha \; T^{-\frac{3}{2}}
$$

$$
BETA = BETA_{TNOM} \left[ \frac{T}{TNOM} \right]^{-\frac{3}{2}} \tag{4.2}
$$

The mobility exponent can be made a user input model parameter instead of using a predefined constant value.

The threshold voltage $VTO$ is also a function of temperature, and an empirical relation of the following form may be used.

$$VTO = VTO_{TNOM} \left[ 1 + TVTO \left( T - TNOM \right) + TVTO2 \left( T - TNOM \right)^2 \right] \tag{4.3}$$

where $TVTO$ and $TVTO2$ are first and second order temperature coefficients respectively.

## 4.7. Model parameter extraction

In the saturation region, the drain current depends only on $v_{gs}$. Therefore, if a very rough estimate of $VTO$ is available, then $i_{ds}$ can be measured as a function of $v_{gs}$ with $v_{ds}$ set to a value just larger than $v_{gs} - VTO$. It is advisable not to set $v_{ds}$ too large in order to avoid the effect of $LAMBDA$ on $i_{ds}$. Under these conditions and if the device is large enough so that effect of $RS$ can be neglected, then:

$$\sqrt{ids} = \sqrt{BETA} \left( v_{gs} - VTO \right) \tag{4.4}$$

This charcateristic is shown graphically in Figure 4.3. $VTO$ is the x-axis intercept of the linear region of this graph having maximum slope. The maximum slope is $\sqrt{BETA}$.

If the effect of $RS$ cannot be neglected, then:

$$v_{gs_{ext}} = VTO + RS \, i_{ds} + \frac{1}{\sqrt{BETA}} \sqrt{i_{ds}} \tag{4.5}$$

where $v_{gs_{ext}}$ is the gate to source voltage between the external gate and source terminals. A nonlinear optimization program can be used to fit the measured data with $v_{gs_{ext}}$ treated as the dependent variable and $\sqrt{i_{ds}}$ and $i_{ds}$ as two independent variables. Initial estimates may be obtained using

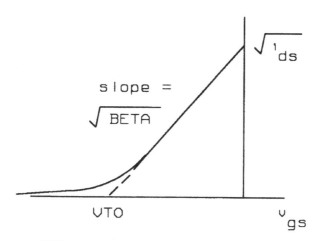

**Figure 4.3:** JFET dc characteristic in saturation.

equation (4.4). To preserve the symmetry of the device, it is assumed that $RD = RS$. The parameter $LAMBDA$ can be determined by measuring $i_{ds}$ for two different values of $v_{ds}$ in saturation, for a constant $v_{gs}$. Under these conditions,

$$\frac{i_{ds_2}}{i_{ds_1}} = \frac{1 + LAMBDA\ v_{ds_2}}{1 + LAMBDA\ v_{ds_1}} \qquad (4.6)$$

and,

$$LAMBDA = \frac{i_{ds_2} - i_{ds_1}}{i_{ds_1}\ v_{ds_2} - i_{ds_2}\ v_{ds_1}} \qquad (4.7)$$

Once good estimates of all the model parameters are obtained, nonlinear optimization can be used to optimize the model parameters using the complete $i_{ds}$ versus $v_{ds}$ characteristics for various values of $v_{gs}$.

# CHAPTER 5

# MOSFET MODELS

## 5.1. Introduction

The metal-oxide-semiconductor field-effect transistor (MOSFET) is the most important device for very-large-scale integrated circuits such as microprocessors and semiconductor memories. MOSFET is also becoming an important power device. Because the current in a MOSFET is transported predominantly by carriers of one polarity only (e.g., electrons in an n-channel device), the MOSFET is referred to as a unipolar device. Although MOSFETs have been made with various semiconductors such as $Ge$, $Si$, $GaAs$, and use various insulators such as $SiO_2$, $Si_3N_4$, and $Al_2O_3$, the most important system is the $Si-SiO_2$ combination [15].

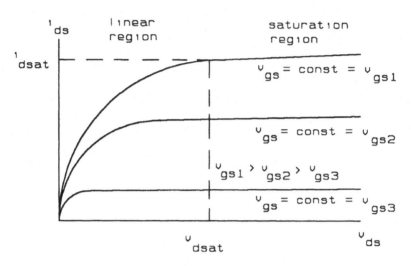

**Figure 5.1:**    Current-voltage characteristics of a long channel MOSFET.

## 5.2. Basic dc characteristics

The MOSFET is a four terminal device and consists of a p-type semiconductor substrate (bulk) into which two $n^+$ regions, the source and the drain, are formed. The metal contact on the insulator is called gate; heavily doped polysilicon or a combination of silicide and polysilicon can also be used as the gate electrode.

We will first consider the basic device characteristics of the long channel MOSFET; that is the channel length is much larger than the sum of the source and drain depletion layer widths [16]. The typical device characteristics are shown in Figure 5.1. Let us consider the situation when a large enough gate voltage is applied to induce an n-type inversion layer between the source and the drain regions, where $Q_n$ is the charge density of electrons per unit surface area in the inversion layer. The magnitude of $Q_n$ dependes on the silicon surface field; hence, it depends on the potential difference between the gate and the inversion layer. For small drain voltages, the channel induced between source and drain essentially behaves like a resistor. As the drain voltage is increased, the average potential difference from gate to the n-type inversion layer will decrease. As a result, $Q_n$ will also decrease and the resistance of the channel will increase. Thus the drain current versus drain voltage characteristic will begin to bend downward from the initial resistor line. As the drain voltage is increased still further, the voltage drop across the oxide near the drain is further reduced until eventually it falls below the level required to maintain an inversion layer. The drain voltage at which this happens is denoted by $v_{dsat}$. At this drain voltage, the channel near the drain disappears. The surface will be merely depleted and no longer inverted. The potential at the end of the inversion layer will be that value for which the gate voltage $v_{gs}$ can still maintain an inversion layer. Once the drain voltage exceeds $v_{dsat}$, the potential at the end of the inversion layer will remain constant, independent of any further increase in the drain voltage, although the point will move somewhat toward the source. The current now is due to the carriers that flow down the inversion layer and are injected into the depletion region near the drain. The magnitude of this current will not change significantly with increasing drain voltage since it depends on the potential drop from the beginning of the inversion layer at the source end to the end of the inversion layer and this potantial drop remains unaltered at $v_{dsat}$. Thus, for drain voltages larger than $v_{dsat}$, the current will not change substantially and will remain at the value $i_{dsat}$. If the gate voltage is increased, the conductance for small values of drain voltage will be larger and the drain voltage at which the current saturates, $v_{dsat}$, will also be larger. As a result, the saturated current, $i_{dsat}$, will also have a larger magnitude. Thus we can distinguish two regions of operation. At low drain voltages, the current-voltage characteristics are nearly ohmic or linear (linear region), while at high drain voltages, the current saturates with increasing drain voltage (saturation region).

The MOSFET equivalent circuit used in most of the circuit simulation programs is shown in Figure 5.2.

### 5.3. Device dimensions

The MOSFET model equations are written in terms of $L_e$ and $W_e$, where $L_e$ is the electrical channel length of the device and $W_e$ is the electrical channel width of the device.

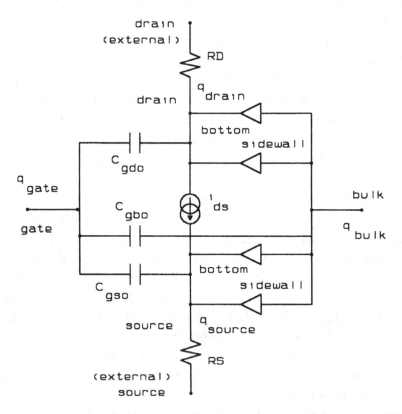

**Figure 5.2:**   MOSFET equivalent circuit.

These electrical dimensions are different from the dimensions that are drawn on the layout. For large devices, these differences are small compared to the device dimensions, but these differences become significant for small size devices. The layout dimensions, denoted by $L$ and $W$, are translated into the optical dimensions $L_0$ and $W_0$, by scaling and biasing operations:

$$L_0 = L \ LMULT + XL \qquad (5.1)$$

$$W_0 = W \ WMULT + XW \qquad (5.2)$$

$LMULT$ and $WMULT$ are the mask scaling factors and $XL$ and $XW$ are the mask biasing factors for the length and width dimensions respectively. These factors depend on the fabrication process. Although the scaling factors may or may not be used in practical fabrication processes, it is convenient to have the freedom of specifying them as model parameters. $L_0$ and $W_0$ are the dimensions of the device on the wafer. These optical dimensions are related to the electrical dimensions by:

$$L_e = L_0 - 2 \ LD \qquad (5.3)$$

$$W_e = W_0 - 2 \ WD \qquad (5.4)$$

where $LD$ and $WD$ are the model parameters affecting the electrical dimensions of the device. $LD$ represents the lateral diffusion of the source and drain regions and $WD$ represents lateral diffusion of the channel stop region and the effect of the formation of the bird's beak. Usually, $LD > 0$ and $WD < 0$, i.e., $L_e < L_0$ and $W_e > W_0$.

It is important to specify these two translations independent of each other. Thus, $W$ and $L$ are specified along with the device specification and $LMULT$, $WMULT$, $XL$, $XW$, $LD$ and $WD$ are specified as model parameters. If both these dimensional translations are specified properly, then it is possible to compute the overlap capacitances from these dimensions, as explained later. SPICE only allows $LD$ as the model parameter. In this case, $L_o$ and $W_o$ must be specified with the device specification. Since it is easier to specify the layout dimensions $L$ and $W$, these are specified with device specification and $LD$ is adjusted so that $L_e = L - 2 LD$. This results in incorrect computation of the parasitic capacitances, if $LD$ is used to compute the overlap capacitances as explained later. For small geometry devices, this can cause a considerable discrepancy between the simulation results and measurements made on the actual fabricated circuits. However, SPICE does *not* use $LD$ to compute the overlap capacitances.

The parameters $LMULT$, $WMULT$, $XL$ and $XD$ are usually determined from the knowledge of the mask making process and the fabrication process; or they can be determined by making optical measurements on different size devices. A relatively simple technique is described below to determine $LD$ [17] and $WD$ [18]. Several modifications and alternatives to these methods have been suggested [19] - [28].

The I-V characteristics of the MOS transistor operating in the linear region can be expressed as:

$$i_{ds} = \mu \; COX \; \frac{W_e}{L_e} \left[ v_{gs} - VTO - \frac{1}{2} v_{ds} \right] v_{ds} \qquad (5.5)$$

where,

   $COX$ = gate oxide capacitance per unit area

   $\mu$ = mobility of carriers in the conducting channel

The intrinsic channel resistance can be written as:

$$R_{chan} = \frac{v_{ds}}{i_{ds}} = A \left( L_o - 2 LD \right) \qquad (5.6)$$

where,

$$A = \frac{1}{\mu \ COX \ W_e \left( v_{gs} - VTO - \frac{1}{2} \ v_{ds} \right)}$$

However, in addition to the intrinsic channel resistance, the MOSFET also consists of extrinsic parasitic resistance $R_{ext}$. Therefore, the measured resistance $R_m$ is given by:

$$R_m = \frac{V_{DS}}{I_{DS}} = R_{ext} + A \left( L_o - 2 \ LD \right) \qquad (5.7)$$

where $V_{DS}$ and $I_{DS}$ are the measured values.

Therefore, if a set of MOSFETs is prepared having different $L_o$'s and same $W_o$, and $R_m$ is plotted as a function of $L_o$ for constant $A$, a straight line would be obtained, as shown in Figure 5.3.

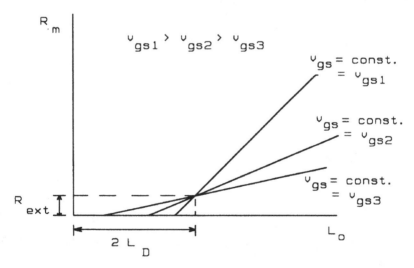

**Figure 5.3:** Measured channel resistance for devices with different channel lengths.

If several lines with different $A$'s were plotted, they would intersect one another at $\left( R_{ext} , 2 \, LD \right)$. Assuming a symmetric device structure, we can obtain $RS = RD = \dfrac{1}{2} R_{ext}$. The coefficient $A$ can be kept constant by using uniform $W_o$ for all the devices. The value of $W_o$ should be relatively large to avoid narrow channel effects. Both $\mu$ and $VTO$ are implicit functions of the effective channel length $L_e$. However, since measurements are made at low $v_{ds}$, usually 0.1 Volt, the dependence of these parameters on $L_e$ is minimized. In addition, application of high gate voltages further minimizes the impact of the variation of $VTO$. But if very high gate voltages are used, then the mobility degradation will affect the extraction of parasitic resistances. It is also assumed that each transistor has same $R_{ext}$. For adjacent transistors on a single chip with the same size of source, drain and contact openings, this assumption is considered reasonable.

The intrinsic channel conductance can be written as:

$$G_{chan} = \frac{i_{ds}}{v_{ds}} = B \left( W_o - 2 \, WD \right) \qquad (5.8)$$

where,

$$B = \frac{1}{L_e} \mu \, COX \left( v_{gs} - VTO - \frac{1}{2} v_{ds} \right)$$

Thus, for a set of devices having same $L_0$, a plot of $G_{chan}$ versus $W_o$ for a constant $v_{gs}$ will be a straight line and $WD$ can be obtained from the intercept of this straight line with the x-axis, as shown in Figure 5.4.

This assumes that the fabricated devices are such that the extrinsic parasitic conductance $G_{ext}$ is much larger than the channel conductance $G_{chan}$. If this is suspected to be an

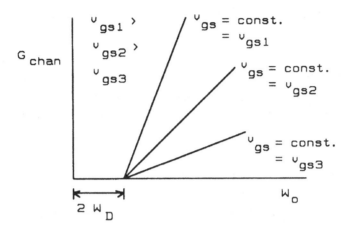

**Figure 5.4:**   Measured channel conductance for devices with different channel widths.

invalid assumption, then $G_{chan}$ can be corrected for $G_{ext}$, since $G_{ext}$ is already determined while computing $LD$.

If the overlap capacitances are specified explicitly, instead of computing from device dimensions, then $W$ and $L$ can be translated directly into $W_e$ and $L_e$, using $WD$ and $LD$. This approach is usally taken in practice since $W$ and $L$ are more readily known compared to $W_o$ and $L_o$. In this case, $W_o$ and $L_o$ in equations (5.5) through (5.8) and the related discussion, are to be replaced by $W$ and $L$ respectively.

## 5.4. MOSFET parasitics

The equivalent circuit of the MOSFET, shown in Figure 5.2, consists of the intrinsic device contributions as well as the parasitic components.

$RS$ and $RD$ are the extrinsic parasitic resistances of the source and drain regions respectively. These resistances can be alternately specified in terms of $RSH$, the sheet resistance of the source and drain regions and $NRS$ and $NRD$, the number of squares of the source and drain regions respectively.

The p-n junctions between the bulk (substrate) and the source and drain regions are represented by parasitic diodes. Each diode is represented as two diodes corresponding to the bottom of the junction and the sidewall or the periphery of the junction. The diode parameters can be specified as absolute values, or they can be specified as per unit junction area for the bottom diodes and per unit junction periphery for the sidewall diodes. The junction area and the periphery for the source and drain junctions are then specified with the MOSFET device information.

The gate terminal has parasitic overlap capacitances with respect to the other three terminals.

$$C_{gso} = CGSO \ W_e \qquad (5.9)$$

$$C_{gdo} = CGDO \ W_e \qquad (5.10)$$

$$C_{gbo} = CGBO \ L_e \qquad (5.11)$$

SPICE does not compute the overlap capacitances if they are not specified. Actually, these can be computed based on other available information:

$$CGSO = COX \ LD \qquad (5.12)$$

$$CGDO = COX \ LD \qquad (5.13)$$

$$CGBO = 2 \ COX \ WD \qquad (5.14)$$

where $COX$ is the oxide capacitance per unit area. This is only a first order approximation and the actual overlap capacitances depend on the shape of the junctions and the fringing fields.

Some circuit simulation programs provide more sophisticated ways of computing the parasitic elements based on
device structure and other geometrical information.

### 5.5. Common model parameters

Some model parameters are related to the basic device
characteristics and are common to many of the MOSFET
models described in this chapter.

The threshold voltage is the gate voltage at the onset of
strong inversion and consists of three components:

$$VTO = vfb + PHI + GAMMA \sqrt{PHI} \qquad (5.15)$$

where,

$GAMMA$ = effect of bulk bias

$PHI$ = surface potential for strong inversion

$vfb$ = flat-band voltage

$VTO$ = threshold voltage with zero bulk bias

$$PHI = 2 \; \frac{k \; TNOM}{q} \; \ln\left(\frac{NSUB}{n_i}\right) \qquad (5.16)$$

$$n_i = 1.45 \times 10^{16} \qquad (5.17)$$

$$COX = \frac{\epsilon_{ox}}{TOX} \qquad (5.18)$$

$$GAMMA = \frac{\sqrt{2 \; q \; \epsilon_{si} \; NSUB}}{COX} \qquad (5.19)$$

$$E_g = 1.16 - \frac{7.02 \times 10^4 \times TNOM^2}{TNOM + 1108.0} \quad (eV) \qquad (5.20)$$

$$\Phi_{ms} = \begin{cases} - \ 0.05 \ - \ 0.5 \ E_g \ - \ 0.5 \ PHI & \text{if } TPG = 1 \\ 0.5 \ E_g \ - \ 0.5 \ PHI & \text{if } TPG = -1 \end{cases}$$

$$\tag{5.21}$$

$$vfb = \Phi_{ms} - \frac{q \ NSS}{COX} \tag{5.22}$$

where,

$k$ = Boltzmann's constant

$q$ = electronic charge

$TNOM$ = nominal temperature in degrees Kelvin

$NSUB$ = doping concentration in the substrate

$n_i$ = intrinsic carrier concentration

$\epsilon_{ox}$ = dielectric constant of oxide

$TOX$ = oxide thickness

$\epsilon_{si}$ = dielectric constant of silicon

$COX$ = oxide capacitance per unit area

$E_g$ = band gap for silicon

$\Phi_{ms}$ = metal-semiconductor work function

$TPG$ = type of gate material

1 => opposite of substrate

-1 => same as substrate

$NSS$ = surface state density in the oxide

The mobility of carriers in the surface inversion layer is specified by $UO$ and two related intermediate quantities are defined below:

$$KP = UO \ COX \tag{5.23}$$

$$\beta = KP \ \frac{W_e}{L_e} \tag{5.24}$$

Many of the above parameters are interrelated and most of the circuit simulation programs provide a flexible way of specifying these parameters.

Some intermediate quantities used by many of the models described in this chapter are defined below:

$$vbi = VTO - GAMMA \ \sqrt{PHI} \tag{5.25}$$
$$= vfb + PHI$$

$$xd = \sqrt{\frac{2 \ \epsilon_{si}}{q \ NSUB}} \tag{5.26}$$

$$sarg = \begin{cases} \sqrt{PHI - v_{bs}} & \text{if } v_{bs} \leq 0.0 \\ \dfrac{\sqrt{PHI}}{1 + 0.5 \ \dfrac{v_{bs}}{PHI} + 0.375 \ \dfrac{v_{bs}^{\ 2}}{PHI^2}} & \text{if } v_{bs} > 0.0 \end{cases} \tag{5.27}$$

$$barg = \begin{cases} \sqrt{PHI - v_{bd}} & \text{if } v_{bd} \leq 0.0 \\ \dfrac{\sqrt{PHI}}{1 + 0.5 \ \dfrac{v_{bd}}{PHI} + 0.375 \ \dfrac{v_{bd}^{\ 2}}{PHI^2}} & \text{if } v_{bd} > 0.0 \end{cases} \tag{5.28}$$

The first equation for $sarg$ cannot be defined for $v_{bs} > PHI$. But during the Newton-Raphson iterations, it is possible to encounter values of $v_{bs}$ greater than *PHI*. Therefore, a smoothing function is used to limit the value of $PHI - v_{bs}$ such that it will always be positive. The smoothing function assures a smooth transition without any discontinuities. A similar technique is used for $v_{bd} > PHI$. SPICE does not use the second order term in the denominator.

## 5.6. Basic device model (level 1)

The **level 1** model in SPICE is the simplest of all available models and represents the basic device characteristics described earlier [29].

$$vth = vbi + GAMMA \ sarg \qquad (5.29)$$

If $v_{ds} \leq v_{gs} - vth$ ( linear region ),

$$i_{ds} = \beta \left( 1 + LAMBDA \ v_{ds} \right) \left[ v_{gs} - vth - \frac{1}{2} v_{ds} \right] v_{ds}$$

$$(5.30)$$

If $v_{ds} > v_{gs} - vth$ ( saturation region ),

$$i_{ds} = \frac{1}{2} \beta \left( 1 + LAMBDA \ v_{ds} \right) \left( v_{gs} - vth \right)^2 \qquad (5.31)$$

## 5.7. Second order effects

The basic device characteristics are derived under the following assumptions:
(1) only drift current is considered;
(2) carrier mobility in inversion layer is constant;
(3) doping in the channel is uniform;
(4) reverse leakage current is negligible;
(5) drain-to-source voltage is small compared to *PHI* in the linear region;

(6) the transverse field in the channel is much larger than the longitudinal field.

The last condition corresponds to the gradual channel approximation.

Since the beginning of the integrated circuit era, the minimum feature length has been reduced by orders of magnitude. As the channel length is reduced, departures from long channel behavior may occur. These departures, the short channel effects, arise as results of a two dimensional potential distribution and high electric fields in the channel region, which make one or more of the above mentioned assumptions invalid.

As the feature size becomes small, the effects of the masking and fabrication processing operations on the device dimensions become comparable to the device size itself, and thus must be taken into account. Also the effects of the parasitic resistances on the device characteristics become significant. These two effects were discussed earlier.

As the channel length is reduced, the depletion layer widths of the source and drain junctions become comparable to the channel length. The potential distribution in the channel now depends on both the transverse field, perpendicular to the current flow and controlled by the gate and the bulk voltages, and the longitudinal field, parallel to the current flow and controlled by the drain voltage. The potential distribution becomes two dimensional and the gradual channel approximation is no longer valid. This two dimensional potential results in degradation of the subthreshold behavior, dependence of the threshold voltage on device dimensions and biasing voltages, and failure of current saturation due to punchthrough.

A reduction in channel length causes an increase in the electric fields in the device. As electric field is increased, the channel mobility becomes field dependent, and eventually velocity saturation occurs. When the field is increased further, carrier multiplication near the drain occurs, leading to substrate current and parasitic bipolar transistor action. High fields also cause hot carrier injection into the oxide, leading to oxide charging and subsequent threshold voltage shift and transconductance degradation.

These second order effects [30] are discussed in some detail in the following sections. Various commonly used MOSFET models are described following this discussion. These models try to incorporate the second order effects by modifying different terms of the basic device model, using theoretical, empirical or semi empirical equations.

Because short channel effects complicate device operation and degrade device performance, these effects should be eliminated or minimized so that a *physical* short channel device can preserve the *electrical* long channel behavior.

### 5.8.  Bulk doping term

The simplified level 1 model assumes that in the linear region, the drain-to-source voltage is small compared to *PHI*. The bulk charge term can be included in the drain current equation as shown below [31]:

$$i_{ds} = \beta \left( 1 + LAMBDA \, v_{ds} \right)$$

$$\left[ \left( v_{gs} - vbi - \frac{1}{2} v_{ds} \right) v_{ds} \right.$$

$$\left. - \frac{2}{3} GAMMA \left( barg^3 - sarg^3 \right) \right] \qquad (5.32)$$

For $v_{ds} \ll PHI - v_{bs}$, this reduces to the simpler equation. Retention of the entire second term or bulk doping term, results in an improved fit to the measured data. The saturation voltage is then given by:

$$v_{dsat} = v_{gs} - vbi + \frac{1}{2} GAMMA^2$$

$$\left[ 1 - \sqrt{1 + \frac{4}{GAMMA^2} (v_{gs} - vbi + PHI - v_{bs})} \right]$$

$$(5.33)$$

To reduce the computational overhead of calculating the bulk doping term, the function,

$$F(v_{ds}, PHI- v_{bs}) = \frac{2}{3} \left[ barg^3 - sarg^3 \right]$$

can be approximated in the range $0 < v_{ds} < 20$ and $0.6 < PHI - v_{bs} < 12.6$ by,

$$F(v_{ds}, PHI- v_{bs}) = sarg \; v_{ds} + \frac{0.25 \; g(PHI - v_{bs})}{sarg} \; v_{ds}^2$$

where,

$$g(PHI - v_{bs}) = 1 - \frac{1}{1.41 + 0.43 (PHI - v_{bs})}$$

## 5.9. Threshold voltage shift

In the basic model, the doping concentration in the substrate is assumed to be constant. In practical devices, however, the doping is generally nonuniform, even for doped substrates that are initially uniform, because the thermal processing causes impurity redistribution. Moreover, in modern MOSFET technology, ion implantation is used extensively to improve device performance. For example, a shallow ion implantation at the $Si- SiO_2$ interface is used for threshold voltage adjustment. The threshold voltage shift is taken into account by simply using the proper value for $VTO$, the zero bias threshold voltage. Another effect of the nonuniform doping concentration is the change in the threshold voltage dependence on substrate bias. The basic model equations suggest that a plot of threshold voltage *vth* versus $\sqrt{PHI - v_{bs}}$ is a straight line with a slope of *GAMMA*, as shown in Figure 5.5. But measurements made on ion implanted devices show a deviation from this straight line behavior. Near zero substrate bias, the slope is higher due to higher doping concentration due to the implant and at higher substrate bias, the slope is lower due to the lower substrate concentration. Some models try to represent this behavior by two straight line segments, while others represent this behavior by a higher order polynomial such as a parabola.

The threshold voltage, *vth*, is strongly influenced by the channel dimensions of small size devices. For short channels, *vth* decreases and for narrow channels it increases [32] - [35]. The integral of the channel doping profile within the

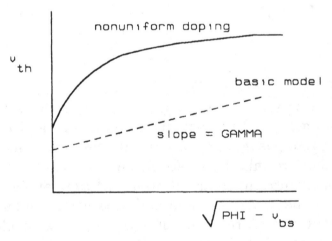

**Figure 5.5:**    Effect of nonuniform substrate doping on threshold voltage.

gate depletion region represents the total depletion charge in the substrate. In the case of short channel devices, this depletion charge decreases from the value predicted by the conventional long channel model, because some of the electric field lines arising from charges near the source and drain are terminated on source and drain respectively, rather than the gate. Therefore, the doping level under the gate is effectively reduced, and the threshold voltage is lower. The charge sharing concept, illustrated in Figure 5.6, splits the depletion layer charge into three regions: region I is controlled by the source, region III is controlled by the drain and region II is controlled by the gate. This decomposition reduces the two dimensional problem to a set of one dimensional problems. A key parameter in the charge sharing model is the charge sharing factor:

$$F_l = \frac{charge\ in\ region\ II}{total\ charge\ in\ regions\ I,\ II\ and\ III} \qquad (5.34)$$

$F_l \approx 1$ for long channel devices and decreases for short channel devices.

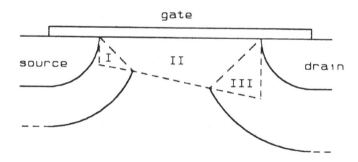

**Figure 5.6:** Charge sharing model for threshold voltage.

In contrast to the threshold voltage decrease for the short channel devices, narrow width devices exhibit an increase of threshold voltage. The depletion boundary under the thin oxide becomes somewhat parabolic and differs from the ideal square boundary predicted by the simple one dimensional model. The thick oxide depletion region on both sides of the gate tends to prevent inversion from occuring under the thin gate oxide near the thin-thick oxide interface. This phenomenon results in a higher threshold voltage compared to the wide width devices, characterized by a narrow width factor $F_w$. This factor can be computed using a charge sharing approach in the width dimension or some empirical relationship may be used:

$$F_w \; \alpha \; \frac{1}{W_e}$$

The threshold voltage for short channel devices also depends on $v_{ds}$ [36]. This effect is known as drain induced barrier lowering and can be modeled as a change in threshold voltage of the form:

$$\Delta vth \; = - \; \frac{ETA}{L_e{}^n} \; v_{ds}$$

where $ETA$ is a user input model parameter and usually $n$ is set to 3.

Thus, the overall threshold voltage is given by,

$$vth \; = vfb \; + \; PHI + F_l \; GAMMA \; \sqrt{PHI - \; v_{bs}}$$
$$+ \; F_w \left( PHI - \; v_{bs} \right) - \; \frac{ETA}{L_e{}^n} \; v_{ds} \qquad (5.35)$$

## 5.10. Mobility reduction

The conduction mechanism between the source and drain strongly depends on the carrier mobility of the charges in the inversion layer [32]. The properties of this thin inversion layer differ from those of the bulk semiconductor mainly because of the large surface-to-volume ratio of the layer. Free carriers are scattered by the surface layer boundary in addition to the scattering mechanisms observed in the semiconductor bulk. The surface properties of the inversion layer have a significant effect on the electrical characteristics of the device. The carrier surface mobility depends on the vertical and the lateral electric fields in the channel. The vertical electric field depends on $v_{gs} - vth$ and the oxide thickness as well as on $v_{bs}$ and bulk doping. The lateral field depends on the channel length and the applied drain voltage. For a wide width device, the bending of the depletion region boundary near the thin-thick oxide interface affects only a very small fraction of the total width. However, when the device width becomes small, this bending affects a large portion of the total width. This corresponds to a large pinch-in field in the transition depletion region along the width direction, i.e. the direction perpendicular to the current flow, and this reduces the carrier mobility. The effective carrier surface mobility can be formulated as:

$$\mu_{eff} = \frac{\mu_0}{1 + f_x\left(E_x\right) + f_y\left(E_y\right) + f_z\left(E_z\right)} \qquad (5.36)$$

where $\mu_0$ is the low field surface mobility, $E$ is the electric field in the channel and $x$, $y$, $z$ are the width, length and vertical dimensions. The above equation assumes that the total carrier scattering probability is equal to the sum of various probabilities due to different scattering mechanisms, i.e., the theory of superposition in scattering probability is assumed to hold. The components of electric field have the following dependencies:

$$E_x \; \alpha \; \frac{1}{W_e}$$

$$E_y \; \alpha \; \frac{1}{L_e} \; , \; v_{ds}$$

$$E_z \; \alpha \; v_{gs} \; - \; vth \; , \; \frac{1}{TOX} \; , \; v_{bs} \; , \; NSUB$$

The functions are generally of the form:

$$f = constant$$

or,

$$f \; \alpha \; \left[ \frac{E}{E_{crit}} \right]^C$$

Sometimes it is more convenient to decouple the effects of various fields as shown below:

$$\mu_{eff} = \frac{\mu_o}{\left[ 1 + f_x(E_x) + f_z(E_z) \right] \left[ 1 + f_y(E_y) \right]}$$

or,

$$\mu_{eff} = \frac{\mu_o}{\left[ 1 + f_x(E_x) \right] \left[ 1 + f_y(E_y) \right] \left[ 1 + f_z(E_z) \right]}$$

## 5.11.  Velocity saturation

When a very small longitudinal field $E_y$ is applied parallel to the current flow, the drift velocity of carriers varies linearly with $E_y$, and the slope is the mobility $\mu$:

$$v = \mu \; E_y$$

The actual drift velocity as a function of the longitudinal field is shown in Figure 5.7 [15]. In short channel devices the electric fields are high enough to reach the region of this curve where the carriers reach their saturation velocity $v_{sat}$. Sometimes this curve is approximated by a straight line up to a critical field $E_{crit}$ and then by a constant drift velocity. Other suggested expressions are [36]:

$$\mu = \frac{\mu_o}{1 + \dfrac{E_y}{E_{crit}}}$$

or,

$$\mu = \frac{\mu_o}{\left[1 + \left(\dfrac{E_y}{E_{crit}}\right)^2\right]^{\frac{1}{2}}}$$

This results in the change in the drain current of the form:

$$i_{ds} \; \alpha \; \frac{1}{1 + \dfrac{v_{ds}}{L_e \, E_{crit}}}$$

As the drain to source voltage $v_{ds}$ is increased, the carrier drift velocity saturates when the critical longitudinal field is reached. In short channel devices, this field is reached only a short distance from the source, resulting in a decrease in $v_{dsat}$ from its long channel value. This decrease is modeled by equations of the form [30] [33]:

$$v_{dsat} = v_{dsat,long} + L_e \, E_{crit} - \sqrt{v_{dsat,long}^2 + L_e \, E_{crit}^2} \tag{5.37}$$

or,

$$v_{dsat} = \frac{v_{gs} - vth}{\alpha} \tag{5.38}$$

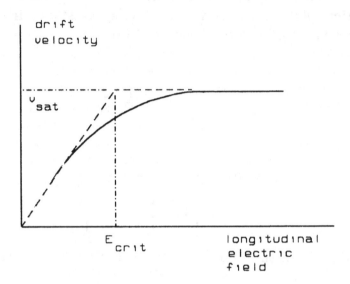

**Figure 5.7:**    Carrier drift velocity versus longitudinal
electric field.

where,

$$\alpha = f \left( v_{gs} \ , \ vth \ , \ v_{bs} \ , \ GAMMA \right)$$

Thus, for short channel devices, velocity saturation occurs at
some low value of $v_{dsat}$, independent of the gate voltage.
The higher the velocity saturation effect, the lower the constant saturation voltage. Above modifications make the
slopes of $i_{ds} - v_{ds}$ curves in the saturation region for different
gate voltages parallel to each other, as is usually observed for
short channel devices. For long channel devices on the
other hand, since $v_{dsat}$ is dependent on the gate voltage, the
saturation slopes are different.

## 5.12. Channel length modulation

The finite output conductance of MOS transistors operating in the saturation region is due to the spreading of the depletion region near the drain which results in a reduction of the channel length [37]. The output conductance is a sensitive function of the oxide thickness under the gate as well as of the impurity concentration in the substrate. A MOS device operates in the saturation region when the drain voltage is increased to a value such that the inversion condition at the end point of the channel near the drain can no longer be maintained by the applied gate voltage. Any further increase in the drain voltage beyond this value then results in the formation of a depleted region of length $\Delta l$ between the drain and the channel leading to a corresponding decrease in the channel length. Assuming that any increase in current beyond the onset of saturation is entirely due to this reduction in channel length, the drain current can be expressed as:

$$i_{ds} = \frac{i_{dsat}}{1 - \dfrac{\Delta l}{L_e}} \qquad (5.39)$$

For long channel devices, $\Delta l \ll L_e$, and the drain current is almost constant in the saturation region. However, for short channel devices, $\Delta l$ may be comparable to the channel length $L_e$ and therefore an appreciable increase in the drain current may be observed. To calculate the variation in drain current, $\Delta l$ must be evaluated as a function of device parameters and applied voltages. The extent of the depletion region depends on the difference between the potential of the drain $v_{ds}$ and the potential at the end of the inversion layer $v_{dsat}$ and on the transverse electric field component near the $Si-SiO_2$ interface, $E_T$.

$$\Delta l = \frac{v_{ds} - v_{dsat}}{E_T} \qquad (5.40)$$

Three contributions to the transverse electric field may be distinguished: (1) the electric field $E_1$ due to the fixed charge in the reverse biased drain junction, (2) the fringing electric field $E_2$ due to the drain-to-gate potential drop, and (3) the fringing electric field $E_3$ due to the potential difference between the gate and the end of the inversion layer.

$$E_1 \ \alpha \ \sqrt{v_{ds} - v_{dsat}} \ , \ \sqrt{NSUB}$$

$$E_2 \ \alpha \ v_{ds} - v_{gs}$$

$$E_3 \ \alpha \ v_{gs} - v_{dsat}$$

## 5.13.  Subthreshold conduction

When gate voltage is below threshold voltage and the semiconductor is in weak inversion, the corresponding drain current is called subthreshold current [15]. The subthreshold region is particularly important for low voltage, low power applications. In weak inversion, the drain current is dominated by carrier diffusion and is an exponential function of the terminal voltages, similar to the currents in a bipolar transistor. The threshold voltage, *vth*, is commonly taken to be the gate to source voltage at which the surface minority carrier concentration at the source is equal to the bulk impurity concentration. This is called *strong* inversion. The drain current, however, does not decrease immediately to zero for $v_{gs} <$ *vth*, so that *vth* does not define an adequate *off* threshold [38]. An alternative definition is the *weak* inversion criterion, which is the gate to source voltage, $v_{off}$, at which the surface minority carrier density at the source is just equal to the intrinsic carrier concentration $n_i$. Some models assume that the channel current is essentially zero for $v_{gs} < v_{off}$ and the subthreshold characteristics are modeled for $v_{off} < v_{gs} < v_{on}$, where $v_{on}$ is defined to be some voltage such that $v_{on} \geq vth$. In the subthreshold region, the surface potential depends on $v_{gs}$ and the assumption that the surface potential is constant at *PHI* is not valid. The subthreshold current is usually modeled as:

$$i_{subth} = I_0 \left( 1 - e^{-\frac{v_{ds}}{v_t}} \right) e^{\left[ \frac{v_{gs} - vth}{n \, v_t} + \frac{v_{ds}}{m \, v_t} \right]} \quad (5.41)$$

where $m$ and $n$ are user input model parameters and $I_0$ can be treated as a user input model parameter or can be computed using theoretical or empirical equations.

$$I_0 = f\left( PHI, \, NSUB, \, TOX \right)$$

For long channel devices, the subthreshold current is independent of the drain voltage for $v_{ds} > 4 \, v_t$, since the surface potential is constant over the entire length of the channel. For short channel devices the surface potential exhibits a localized potential barrier. The height of this barrier is reduced by increasing the drain voltage and the position of the peak shifts closer to the source region. Therefore, the subthreshold current depends on the drain voltage for a larger $v_{ds}$ range.

Circuit simulation models were first developed for the above threshold operations since this was considered the most important region of operation. Extending the model to the subthreshold region gives rise to a problem that the surface potential approximation used in the strong inversion region is not valid in the weak inversion region. A discontinuity exists between the strong inversion or the above threshold region and the weak inversion or the subthreshold region. Drain current is modeled as pure diffusion current in the subthreshold region, whereas it is modeled as pure drift current in the above threshold region. Some empirical method is usually used to provide a smooth transition between the two regions [39] - [41].

### 5.14. Avalanche current

When the drain voltage becomes sufficiently high so that a depleted pich-off region is formed at the drain end, an avalanche can occur within the pinch-off region. From the avalanche plasma, the generated electrons enter the drain and the generated holes are collected by the substrate terminal and constitute the substrate current. The requirements of noise margins and process margins have made it difficult to scale the operating voltages with the decrease of feature size. Consequently, the electric field at the drain increases for smaller geometries. The impact ionization coefficient is an exponential function of this electric field. For short channel devices, an additional effect is caused by the avalanche generated hole current [15]. As the source-drain separation is reduced, some hole current can flow to the source. A substantial substrate current can make the source-bulk junction forward biased causing electron injection into the substrate. This injection leads to a parasitic n-p-n (source-substrate-drain) bipolar transistor action.

The impact ionization current can cause several undesirable effects [33]. The primary impact ionization results in a majority carrier substrate current which causes local substrate debiasing and results in circuit failure for high speed circuits. A secondary impact ionization occurs when the majority carriers are accelerated through the drain to substrate field. This results in a nonthermal minority carrier leakage in the substrate, resulting in a refresh problem for dynamic memories.

Because impact ionization substrate currents are extremely sensitive to device dimensions, channel doping and terminal voltages, the magnitude of such currents varies greatly with the circuit and the processed wafer [42]. Circuit

designers need to match their back bias circuits to the substrate currents expected during actual circuit operation. To
do this, dc substrate current must be modeled in circuit
simulation tools. The substrate current is of the form [43]:

$$i_{sub} \propto i_{ds} , v_{ds} - v_{dsat} , e^{-\frac{constant}{v_{ds} - v_{dsat}}} \qquad (5.42)$$

This will appear as a current source in the equivalent circuit
of Figure 5.2, flowing from the drain to the bulk terminal.

The avalanche breakdown determines the maximum
drain voltage that can be applied to the device. Very small
source-to-drain spacings cause another mechanism called
*punchthrough* [44], which also limits the maximum usable
drain voltage. Punchthrough results when the drain depletion region edge touches the source depletion edge.

## 5.15. Oxide charging

Another high filed effect related to the avalanche
current mechanism is oxide charging [15] [45]. As the field
along the channel becomes high, some electrons in the
inversion layer can gain sufficient energy to surmount the
$Si-SiO_2$ energy barrier (3.1 eV) and get injected into the
gate oxide. Hot electrons can also be injected from the
avalanche plasma formed near the drain region. Thermally
generated carriers can also be injected into the oxide due to
a large transverse field in the bulk semiconductor. In addition to giving rise to a gate current, the injected hot electrons cause the threshold voltage to shift toward more positive voltage. The transconductance, slope of $i_{ds}$ versus $v_{gs}$
curve, becomes smaller due to reduced channel mobility.
The subthreshold current becomes larger because of the
increased interface trap density.

Long term operation of the device is seriously affected
by oxide charging, because the charging continues to

increase with time during device operation. As a result of this cumulative degradation, oxide charging limits the maximum voltage levels that can be applied for a given specific device lifetime.

## 5.16.  Device symmetry

All the models described in this chapter assume that $v_{ds} \geq 0$, i.e. the drain is at a higher potential than the source. If this is not the case, then the roles of the source and the drain are reversed. Thus, for model computations, among the topological source and drain terminals, the terminal which is at a lower potential is to be treated as the source. The topological connectivity is used to compute the terminal voltages and also to load the currents and charges in the right hand side vector and the jacobians into the admittance matrix. The topological source and drain may become electrical darin and source if the device is in inverse mode of operation. The mapping between the topological and the electrical connections must be done properly to avoid any potential nonconvergence problems.

If the fabrication process is such that the device structure is not symmetric, then the model equations have to be implemented in such a way that the drain current can be computed for $v_{ds} < 0$.

## 5.17.  Voltage limiting

The concept of voltage limiting was explained earlier. The bulk-drain and bulk-source diode voltages are limited similar to the junction voltage limiting for diodes. SPICE also uses branch voltage limiting for gate-source and drain-source voltages. These equations are completely empirical and are based on heuristics. Some other simulators do not use limiting or use much simpler limiting equations without encountering any numerical problems.

### 5.17.1. Gate-source voltage

$v_{gso}$ is the old value of $v_{gs}$. The solution of the circuit yields $\hat{v}_{gs}$ for use in the next iteration. It is desired to choose $v_{gs1} \leq \hat{v}_{gs}$. Following intermediate quantities are used:

$$\Delta v_{gs} = \left| \hat{v}_{gs} - v_{gso} \right|$$

$$VTO_x = VTO + 3.5$$

$$v_{hi} = 2 \left| v_{gso} - VTO \right| + 2$$

$$v_{lo} = \left| v_{gso} - VTO \right| + 3$$

where $VTO$ is the threshold voltage.

If $v_{gso} < VTO$ ( *off* region) and $\hat{v}_{gs} > v_{gso}$ ($v_{gs}$ increasing):

$$v_{gs1} = \begin{cases} VTO + 0.5 & \text{if } \hat{v}_{gs} > VTO + 0.5 \\ v_{gso} + v_{lo} & \text{if } \Delta v_{gs} > v_{lo} \\ \hat{v}_{gs} & \text{if } \Delta v_{gs} \leq v_{lo} \end{cases}$$

If $v_{gso} < VTO$ ( *off* region) and $\hat{v}_{gs} \leq v_{gso}$ ($v_{gs}$ decreasing):

$$v_{gs1} = \begin{cases} v_{gso} - v_{hi} & \Delta v_{gs} > v_{hi} \\ \hat{v}_{gs} & \Delta v_{gs} \leq v_{hi} \end{cases}$$

If $v_{gso} < VTO_x$ ( *intermediate* region):

$$v_{gs1} = \begin{cases} \min\left( \hat{v}_{gs}, VTO + 4 \right) & \text{if } \hat{v}_{gs} > v_{gso} \\ \max\left( \hat{v}_{gs}, VTO - 0.5 \right) & \text{if } \hat{v}_{gs} \leq v_{gso} \end{cases}$$

If $v_{gso} \geq VTO_x$ ( *on* region):

$$
v_{gs1} = \begin{cases}
v_{gso} + v_{hi} & \text{if } \hat{v}_{gs} > v_{gso} \\
\max \left( \hat{v}_{gs}, VTO + 2 \right) & \text{if } \hat{v}_{gs} < VTO_x \\
v_{gso} - v_{lo} & \text{if } \hat{v}_{gs} > v_{lo} \\
\hat{v}_{gs} & \text{if } \hat{v}_{gs} \leq v_{lo}
\end{cases}
$$

## 5.17.2. Drain-source voltage

$v_{dso}$ is the old value of $v_{ds}$. The solution of the circuit yields $\hat{v}_{ds}$ for use in the next iteration. It is desired to choose $v_{ds1} \leq \hat{v}_{ds}$.

$$
v_{ds1} = \begin{cases}
\min \left( \hat{v}_{ds}, 4 \right) & \text{if } v_{dso} < 3.5, \hat{v}_{ds} > v_{dso} \\
\max \left( \hat{v}_{ds}, -0.5 \right) & \text{if } v_{dso} < 3.5, \hat{v}_{ds} \leq v_{dso} \\
\min \left( \hat{v}_{ds}, 3 v_{dso} + 2 \right) & \text{if } v_{dso} \geq 3.5, \hat{v}_{ds} > v_{dso} \\
\max \left( \hat{v}_{ds}, 2 \right) & \text{if } v_{dso} \geq 3.5, \hat{v}_{ds} \leq v_{dso}, \hat{v}_{ds} < 3.5 \\
 & \text{if } v_{dso} < 3.5, \hat{v}_{ds} > v_{dso}, \hat{v}_{ds} \geq 3.5 \\
\hat{v}_{ds} &
\end{cases}
$$

SPICE does not do the voltage limiting correctly when $v_{ds} < 0$.

## 5.18. Geometry dependence

Ideally we would like to use only one set of model parameters to cover the entire spectrum of possible device sizes. In VLSI design, a wide range of device sizes is encountered and none of the available models can model the device characteristics accurately for devices having such a wide geometry variation [46] [47]. Of course, some models can cover a wider range than others. To maintain model accuracy over the entire range of device sizes used in a circuit, it is generally advisable to divide the range of device sizes in *geometry bins*, as shown in Figure 5.8. The length spectrum of interest is from $L_{min}$ to $L_{max}$ and the width spectrum of interest is $W_{min}$ to $W_{max}$. One set of model parameters is used for each bin. Some models can cover a wider range of device sizes with a single set of model parameters and thus will require fewer bins. It is necessary to have more bins to cover the short channel and narrow width range since the device characteristics may not scale well with device sizes in that range. The range of lengths and widths covered by each bin can progressively increase for larger size devices, where the device characteristics may scale well with device sizes. Generally, device characteristics scale better in the width dimension and hence fewer bins are required in the width dimension compared to the length dimension.

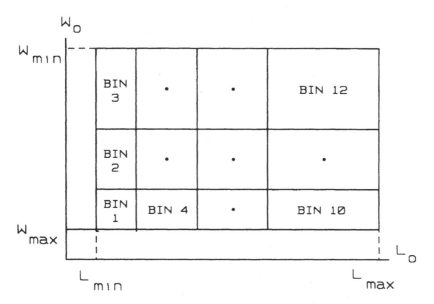

**Figure 5.8:** Geometry bins covering the device sizes of interest.

## 5.19. Level 2 model

The **level 2** model in SPICE is an analytical one dimensional model which incorporates many of the second order effects for small size devices [48]. This model can accurately cover a reasonable range of device sizes, but is computationally quite complex.

### 5.19.1. Threshold voltage

The threshold voltage equation uses the charge sharing model to compute the effect of short channel length on the threshold voltage. The equations also include the narrow width effect. The voltage $v_{on}$ is defined for modeling the subthreshold region.

$$F_N = \frac{1}{8} \frac{2 \pi \epsilon_{si} DELTA}{COX \ W_e}$$

$$argss = \frac{1}{2} \frac{XJ}{L_e} \left[ \sqrt{1 + 2 \frac{xd \ sarg}{XJ}} - 1 \right]$$

$$argsd = \frac{1}{2} \frac{XJ}{L_e} \left[ \sqrt{1 + 2 \frac{xd \ barg}{XJ}} - 1 \right]$$

$$\gamma_{SD} = GAMMA \ ( \ 1 - \ argss - \ argsd \ )$$

$$cfs = q \ NFS$$

$$Q_{dep} = COX \ \gamma_{SD} \ sarg$$

$$cd = \frac{\partial Q_{dep}}{\partial v_{bs}}$$

$$xn = 1 + \frac{cfs}{COX} + \frac{cd}{COX} + F_N$$

$$vth = vbi + F_N \ ( \ PHI - \ v_{bs} \ ) + \gamma_{SD} \ sarg$$

$$vbin = vbi + F_N \ (\ PHI - \ v_{bs}\ )$$

$$v_{on} = vth + \frac{k\ T}{q}\ xn$$

### 5.19.2. Mobility reduction

The mobility reduction equation is a power law relationship accounting for mobility reduction due to the vertical electric field due to the gate voltage. The mobility starts decreasing when the vertical electric field increases beyond a critical field $UCRIT$.

$$\mu_{fact} = \left[\frac{\epsilon_{si}\ UCRIT}{COX\ (\ v_{gs} - \ v_{on} - \ UTRA\ v_{ds}\ )}\right]^{UEXP}$$

$$\mu_{eff} = \mu_{fact}\ UO$$

### 5.19.3. Velocity saturation

If the maximum carrier velocity saturation $VMAX$ is not specified, then SPICE computes the saturation voltage $v_{dsat}$ assuming that the channel pinches off at the drain. With the correction for small size effects,

$$\eta = 1 + F_N$$

$$\gamma_D = \frac{\gamma_{SD}}{\eta}$$

$$vgsx = \begin{cases} v_{gs} & \text{if } NFS = 0.0 \\ v_{gs} & \text{if } NFS > 0.0 \text{ and } v_{gs} \geq v_{on} \\ v_{on} & \text{if } NFS > 0.0 \text{ and } v_{gs} < v_{on} \end{cases}$$

$$v_{dsat} = \frac{vgsx - vbin}{\eta}$$

$$+ \frac{1}{2} \gamma_D{}^2 \left[ 1 - \sqrt{1 + \frac{4}{\gamma_D{}^2} \left( \frac{vgsx - vbin}{\eta} + PHI - v_{bs} \right)} \right]$$

If carrier velocity saturation is taken into account, $v_{dsat}$ is given by,

$$VMAX = \frac{\mu_{eff}\ I}{L_e \left[ v_{gs} - vbin - \eta\ v_{dsat} - \gamma_{SD} \sqrt{PHI - (v_{bs} - v_{dsat})} \right]}$$

where,

$$I = \left\{ \left( v_{gs} - vbin - \frac{1}{2}\ \eta\ v_{dsat} \right) v_{dsat} \right.$$

$$\left. - \frac{2}{3} \gamma_{SD} \left[ (PHI - (v_{bs} - v_{dsat}))^{\frac{3}{2}} - (PHI - v_{bs})^{\frac{3}{2}} \right] \right\}$$

This equation is transformed into a quartic equation as follows:

$$xv = \frac{VMAX\ L_e}{\mu_{eff}}$$

$$x = \sqrt{v_{dsat} + PHI - v_{bs}}$$

$$v_1 = \frac{v_{gs} - vbin}{\eta} + PHI - v_{bs}$$

$$v_2 = PHI - v_{bs}$$

$$xv = \frac{\left( v_1 - \frac{v_2}{2} - \frac{x^2}{2} \right) (x^2 - v_2) - \frac{2}{3}\ \gamma_D \left( x^3 - v_2{}^{\frac{3}{2}} \right)}{v_1 - \gamma_D\ x - x^2}$$

$$A = \frac{4}{3} \gamma_D$$

$$B = -2\left(v_1 + xv\right)$$

$$C = -2\gamma_D \, xv$$

$$D = 2v_1\left(v_2 + xv\right) - v_2^2 - \frac{4}{3}\gamma_D \, v_2^{\frac{3}{2}}$$

$$x^4 + A\,x^3 + B\,x^2 + C\,x + D = 0$$

This equation is solved by Ferrari's method. Among the 2 or 4 real roots, the smallest positive root is the valid solution.

### 5.19.4. Channel length modulation

If $LAMBDA > 0.0$,

$$l_{fact} = 1 - LAMBDA \, v_{ds}$$

If $LAMBDA \leq 0.0$ and $NSUB > 0.0$ and $VMAX = 0.0$,

$$l_{fact} = 1 - \frac{xd}{L_e}\left[\frac{\left(v_{ds} - v_{dsat}\right)}{4} + \sqrt{1 + \left[\frac{v_{ds} - v_{dsat}}{4}\right]^2}\right]^{\frac{1}{2}}$$

This formulation does not include the effect of the field between gate and drain and gate and pich-off point respectively, but insures the continuity of current and its first derivative at the transition from linear to saturation region. The above equation tends to overestimate the conductance in the saturation region.

If saturation velocity is specified, i.e. if $LAMBDA \leq 0.0$ and $NSUB > 0.0$ and $VMAX > 0.0$,

$$l_{fact} = 1 - \frac{xd}{\sqrt{NEFF} \; L_e} \times$$

$$\left[ \sqrt{\left( \frac{VMAX \; xd}{2\sqrt{NEFF} \; \mu_{eff}} \right)^2 + v_{ds} - v_{dsat}} - \frac{VMAX \; xd}{2\sqrt{NEFF} \; \mu_{eff}} \right]$$

The punchthrough mechanism is not modeled, but the channel length is prevented from going negative:

$$xwb = xd \; \sqrt{PB}$$

$$l_{eff} = L_e \; l_{fact}$$

$$l_{eff} = \frac{xwb}{1 + \dfrac{xwb - l_{eff}}{xwb}} \qquad \text{if } l_{eff} < xwb$$

$$l_{fact} = \frac{l_{eff}}{L_e}$$

### 5.19.5. Drain current

$$\beta_{eff} = \beta \; \frac{\mu_{fact}}{l_{fact}}$$

$$vdsx = \begin{cases} v_{ds} & v_{ds} \leq v_{dsat} \\ v_{dsat} & v_{ds} > v_{dsat} \end{cases}$$

$$i_{ds} = \beta_{eff} \left\{ \left( vgsx - vbin - \frac{1}{2} \eta \; vdsx \right) \right.$$

$$\left. - \frac{2}{3} \gamma_{SD} \left[ (PHI - (v_{bs} - vdsx))^{\frac{3}{2}} - (PHI - v_{bs})^{\frac{3}{2}} \right] \right\}$$

### 5.19.6. Subthreshold conduction

A voltage $v_{on} > vth$ is defined, which marks the transition from weak to strong inversion characteristc. If $v_{gs} \leq v_{on}$ and $NFS > 0.0$,

$$i_{ds} = i_{ds} \times e^{\frac{q}{kT} \frac{v_{gs} - v_{on}}{xn}}$$

This equation insures the continuity of the drain current at $v_{on}$ but does not insure the continuity of the first derivative.

## 5.20. Level 3 model

The **level 3** model is a semi empirical model described by a set of parameters defined by curve fitting rather than physical background [48] [49]. This model addresses the issue of computational efficiency. The model computation time for this model is about one third to one half of the computation time for the level 2 model. However, this model can cover only a limited range of device sizes, necessitating more geometry bins to cover the device size range of interest.

### 5.20.1. Threshold voltage

Threshold voltage dependence on channel length is modeled using the charge sharing approach. The channel width dependence is modeled by an empirical term. Drain induced barrier lowering is also included. All the three effects are decoupled by having separate correction terms.

$$F_N = \frac{1}{4} \frac{2 \pi \epsilon_{si} \ DELTA}{COX \ W_e}$$

$$\sigma = \frac{\Omega \ ETA}{COX \ W_e{}^3}$$

$$\Omega = 8.15 \times 10^{-22}$$

$$W_p = xd \ sarg$$

$$d_0 = 0.0631353$$

$$d_1 = 0.8013292$$

$$d_2 = -0.01110777$$

$$\frac{W_c}{XJ} = d_0 + d_1 \frac{W_p}{XJ} + d_2 \left[\frac{W_p}{XJ}\right]^2$$

$$F_S = 1 - \frac{XJ}{L_e} \left\{ \frac{LD + W_c}{XJ} \left[ 1 - \left( \frac{\frac{W_p}{XJ}}{1 + \frac{W_p}{XJ}} \right)^2 \right]^{\frac{1}{2}} - \frac{LD}{XJ} \right\}$$

$$xn = 1 + \frac{q \ NFS}{COX} + \frac{GAMMA \ F_S \ sarg + F_N \left( PHI - v_{bs} \right)}{2 \left( PHI - v_{bs} \right)}$$

$$vth = vfb + PHI - \sigma \ v_{ds} + GAMMA \ F_S \ sarg$$
$$+ F_N \left( PHI - v_{bs} \right)$$

$$v_{on} = vth + \frac{k \ T}{q} \ xn$$

## 5.20.2. Mobility reduction

A simple empirical equation is used to model the mobility reduction due to the vertical electric field caused by the gate voltage.

$$vgsx = \max \left( v_{gs} , v_{on} \right)$$

$$\mu_{fact} = \frac{1}{1 + THETA\ \left( vgsx - v_{on} \right)}$$

$$\mu_S = \mu_{fact}\ UO$$

### 5.20.3. Velocity saturation

The velocity saturation is mdeled by using an effective mobility which is a function of $v_{ds}$,

$$F_{drain} = \frac{1}{1 + \dfrac{\mu_S}{VMAX\ W_e}\ vdsx}$$

The saturation voltage is given by,

$$F_B = \frac{GAMMA\ F_S}{4\ sarg} + F_N$$

$$v_{dsat} = \frac{vgsx - v_{on}}{1 + F_B} + \frac{VMAX\ W_e}{\mu_S}$$
$$- \sqrt{\left( \frac{vgsx - v_{on}}{1 + F_B} \right)^2 + \left( \frac{VMAX\ W_e}{\mu_S} \right)^2}$$

$$vdsx = \min\ \left( v_{ds}\ ,\ v_{dsat} \right)$$

### 5.20.4. Drain current

$$i_{ds} = \beta\ \mu_{fact}\ F_{drain}\ \left( vgsx - v_{on} - \frac{1 + F_B}{2}\ vdsx \right) vdsx$$

### 5.20.5. Channel length modulation

If $v_{ds} > v_{dsat}$ then the channel length modulation factor is computed. If $VMAX = 0.0$,

$$\Delta l = xd\ \sqrt{KAPPA\ \left( v_{ds} - v_{dsat} \right)}$$

If $VMAX > 0.0$,

$$i_{dsat} = i_{ds}$$

$$g_{dsat} = i_{dsat} \left( 1 - F_{drain} \right) \frac{\mu \, S}{W_e \, VMAX}$$

$$E_p = \frac{i_{dsat}}{g_{dsat} \, W_e}$$

$$\Delta l = \sqrt{\left( \frac{E_p \, xd^2}{2} \right)^2 + KAPPA \, xd^2 \left( v_{ds} - v_{dsat} \right)}$$

$$- \frac{xd^2 \, E_p}{2}$$

The voltage across the depletion region of length $\Delta l$ is assumed to be less than $v_{ds} - v_{dsat}$ by a factor $KAPPA$. The channel length modulation is limited using:

If $\Delta l > \frac{1}{2} W_e$,

$$\Delta l = W_e - \frac{W_e{}^2}{4 \, \Delta l}$$

$$l_{fact} = \frac{1}{1 - \dfrac{\Delta l}{W_e}}$$

$$i_{ds} = i_{ds} \times l_{fact}$$

The output conductance, slope of the $i_{ds}$ versus $v_{ds}$ characteristic, shows a minimum at $v_{ds} - v_{dsat}$. Initially the conductance decreases as $v_{ds}$ is increased. As $v_{ds}$ reaches $v_{dsat}$, the conductance reaches a minimum. As $v_{ds}$ is increased beyond $v_{dsat}$, the output conductance increases slightly and then remains almost constant. This is contrary to the monotonically decreasing behavior in practical devices.

### 5.20.6. Subthreshold conduction

The subthreshold conduction equations are the same as used in the level 2 model:

$$i_{ds} = i_{ds} \times e^{\frac{q}{k\,T}\frac{v_{gs} - v_{on}}{xn}}$$

### 5.21. CSIM model

The **CSIM** (Compact Short-channel IGFET Model) model was developed at AT&T Bell Laboratories. This is a practical model with emphasis on the simplicity of its formulation [50] [51]. The model simplicity allows efficient extraction of reliable model parameters resulting in accurate model fitting of measured data. The formulation of earlier models, like the level 2 model, use complex equations for drain current, involving long algebraic expressions and computation of a large number of square roots. The models are further complicated by addition of short channel effects and subthreshold conduction. The price paid in using the complex models is longer computation time. For the CSIM model, the computation time is reduced by about 50 per cent compared to the level 2 model.

$$vth = VTO + K1\left(\sqrt{PHI - v_{bs}} - \sqrt{PHI}\right)$$

*Off* region $(v_{gs} < vth - ETA\,v_{ds})$:

$$i_{ds} = 0$$

*Linear* region $\left(0 < v_{ds} < \dfrac{v_{gs} - vth}{a - ETA}\right)$:

$$i_{ds} = \beta_{eff}\left[(v_{gs} - vth)\,v_{ds} - \left(\frac{a}{2} - ETA\right)v_{ds}^{2}\right]$$

$$\beta_{eff} = \frac{\beta}{1 + UF\left(v_{gs} - vth + ETA\,v_{ds}\right)}$$

$$a = 1 + \frac{0.5\,K1\,g}{\sqrt{PHI-\,v_{bs}}}$$

$$g = 1 - \frac{1}{1.744 + 0.8364\left(PHI-\,v_{bs}\right)}$$

*Saturation* region $\left(v_{ds} \geq \dfrac{v_{gs} - vth}{a - ETA}\right)$:

$$i_{ds} = \frac{\beta_{eff}}{2}\left(v_{gs} - vth + ETA\,v_{ds}\right)^2$$

*VTO* is the threshold voltage *vth* at zero $v_{bs}$ and $K1$ is the slope of the *vth* versus $\sqrt{PHI - v_{bs}}$ curve. *UF* is the mobility degradation factor and *ETA* describes the finite output conductance in saturation. The factor "*a*" describes the lowering of drain current due to bulk charge in the substrate. When *a* is one and *ETA* and *UF* are set to zero, the equations reduce to the basic model described earlier.

This model can accurately reproduce the device characteristics of short channel devices. However, the parameters $K1$, *UF* and *ETA* are functions of $\dfrac{1}{L_e}$ and hence one set of model parameters can cover only a limited range of channel length values. This range can be increased by making the model parameters to be functions of device geometry similar to the BSIM model described below.

## 5.22. BSIM model

The **BSIM** (Berkeley Short-channel IGFET Model) model [51] [52] builds upon the CSIM model with many enhancements. Since a fully physics oriented modeling approach usually makes parameter extraction particularly

difficult, a semi empirical approach is adopted in developing
BSIM to cope with the rapid advances in technology and to
make automated parameter extraction possible. This model
has many model parameters that are bias dependent and if
care is not taken in implementing the model, nonphysical
model parameter values may result. Also, the model has a
large number of empirical parameters. This has two conse-
quences. Firstly, model parameter values need to be
specified properly. It is possible to specify peculiar combina-
tions of model parameter values, which will result in nonphysical device characteristics. Secondly, the use of an
automated extraction procedure is required to obtain model
parameter values.

### 5.22.1.  Threshold voltage

$$\eta = ETA + X2E\ v_{bs} + X3E\ \left(\ v_{ds} -\ VDD\ \right)$$

$$vth = VFB + PHI + K1\ \left(\ \sqrt{PHI - v_{bs}}\ \right)$$

$$- K2\ \left(\ PHI -\ v_{bs}\ \right) -\ \eta\ v_{ds}$$

The $K1$ and $K2$ terms together model the nonuniform doping effect. $\eta$ accounts for the drain induced barrier lowering
as well as for finite conductance in the saturation region.
$VDD$ is the drain voltage at which the saturation region
measurements are made. For $v_{ds} > VDD$, $\eta$ may become
negative, resulting in an *increase* in $vth$. Also, for large $v_{bs}$,
the $K2$ term may dominate the $K1$ term, giving rise to a
*reduction* in $vth$. Both these effects are nonphysical.

### 5.22.2.  Mobility reduction

Mobility degradation due to vertical and lateral fields is
modeled.

$$U_{gs} = U0 + X2U0\ v_{bs}$$

$$U_{ds} = \frac{U1 + X2U1\ v_{bs} + X3U1\left(v_{ds} - VDD\right)}{L_e}$$

### 5.22.3. Effective beta

The effective beta, the gain parameter, is obtained by quadratic interpolation through three data points: the mobility at $v_{ds} = 0$, the mobility at $v_{ds} = VDD$ and the slope of the mobility with respect to $v_{ds}$ at $v_{ds} = VDD$. Since these three values are specified independent of eachother, it is possible to get negative drain conductance in saturation if the values are not specified properly.

$$\beta_{0_0} = MUZ\ COX\ \frac{W_e}{L_e}$$

$$\beta_{0_b} = X2MZ\ COX\ \frac{W_e}{L_e}$$

$$\beta\Big|_{vds\,=0} = \beta_{0_0} + \beta_{0_b}\ v_{bs}$$

$$\beta_{VDD} = MUS\ COX\ \frac{W_e}{L_e}$$

$$\beta_{VDD_b} = X2MS\ COX\ \frac{W_e}{L_e}$$

$$\beta\Big|_{vds\,=\,VDD} = \beta_{VDD} + \beta_{VDD_b}\ v_{bs}$$

$$\beta_{VDD_d} = X3MS\ COX\ \frac{W_e}{L_e}$$

$$\frac{\partial\beta}{\partial vds}\Big|_{vds\,=\,VDD} = \beta_{VDD_d}$$

If $v_{ds} > VDD$,

$$\beta_0 = \beta \bigg|ds = VDD + \frac{\partial \beta}{\partial vds}\bigg|_{v_{ds} = VDD} \left( v_{ds} - VDD \right)$$

If $v_{ds} \leq VDD$,

$$\beta_0 = \beta \bigg|_{vds = 0}$$

$$+ v_{ds} \left\{ \left[ \frac{2 \left( \beta_{vds = VDD} - \beta_{vds = 0} \right)}{VDD} - \frac{\partial \beta}{\partial vds}\bigg|_{vds = VDD} \right] \right.$$

$$+ \left. \left[ \frac{- \beta_{vds = VDD} + \beta_{vds = 0} + \frac{\partial \beta}{\partial vds}\bigg|_{vds = VDD} \quad VDD}{VDD^2} v_{ds} \right] \right\}$$

$$\beta_{eff} = \frac{\beta_0}{1 + U_{gs} \left( v_{gs} - vth \right)}$$

### 5.22.4. Saturation Voltage

$$g = 1 - \frac{1}{1.744 + 0.8364 \left( PHI - v_{bs} \right)}$$

$$a = 1 + \frac{g \, K1}{2 \, \sqrt{PHI - v_{bs}}}$$

$$vc = \frac{U_{ds} * \left( v_{gs} - vth \right)}{a}$$

$$k = \frac{1 + vc + \sqrt{1 + 2 \, vc}}{2}$$

$$v_{dsat} = \frac{v_{gs} - vth}{a \, \sqrt{k}}$$

## 5.22.5. Drain Current

$$vz = \begin{cases} v_{ds} & \text{if } v_{ds} \leq v_{dsat} \\ v_{dsat} & \text{if } v_{ds} > v_{dsat} \end{cases}$$

$$i_{ds} = \frac{\beta_{eff}}{\left(1 + U_{ds} \ vz\right)} \left[ (v_{gs} - vth) \ vz - \frac{a}{2} \ vz^2 \right]$$

## 5.22.6. Subthreshold Conduction

$$n = N0 + NB \ v_{bs} + ND \ v_{ds}$$

$$v_t = \frac{k \ T}{q}$$

$$i_{exp} = \beta_{0_0} \ v_t^2 \ e^{1.8} \ e^{\frac{v_{gs} - vth}{n \ v_t}} \left( 1 - e^{-\frac{v_{ds}}{v_t}} \right)$$

$$i_{limit} = 4.5 \ \beta_{0_0} \ v_t^2$$

$$i_{subth} = \frac{i_{limit} \ i_{exp}}{i_{limit} + i_{exp}}$$

$$i_{ds} = i_{ds} + i_{subth}$$

## 5.22.7. Geomtery dependence

Each model parameter has three components:

$$P = P_0 + \frac{P_L}{L_e} + \frac{P_W}{W_e}$$

$P_0$ is the offset value, $P_L$ is the channel length sensitivity and $P_W$ is the channel width sensitivity.

In practice, it is observed that some of the model parameters do not obey the inverse length or width dependence. This limits the range of device dimensions that can be covered by one set of model parameters. Also, the parameter component $P_0$ corresponds to the value of the model parameter for a device having very long channel length $L_\infty$ and very wide channel width $W_\infty$. Thus, implicitly the model parameter values are scaled from $L_\infty$ to $L_e$ and from $W_\infty$ to $W_e$, which is a large dynamic range. Better accuracy can be obtained by choosing the most commonly used device as the reference device $(L_R , W_R)$ and then scale the model parameters with respect to this device.

$$P = P_R + P_L \left( \frac{1}{L_e} - \frac{1}{L_R} \right) + P_W \left( \frac{1}{W_e} - \frac{1}{W_R} \right)$$

where $P_R$ is now the model parameter value for the reference device. This modification can insure the model accuracy over a wider geometry range, reducing the number of geometry bins required to cover the desired geometry range.

## 5.23. Table lookup models

Table lookup models represent the device characteristics in terms of tables of data points instead of using analytical equations. These data points can be obtained from measurements made on test devices or they can be obtained from device analysis programs if fabricated devices are not available. Several approaches [53] - [60] can be used to implement the lookup tables and an approach proposed in [61] is described here.

In the simulation of large circuits, a major portion of the simulation time is spent in evaluating the transistor models. With the use of table models the model evaluation time can be reduced considerably, while maintaining good accuracy in terms of individual device charcateristics as well as overall circuit behavior. A difficulty in using analytical models is that, as the technology progresses, the present models may not be adequate to characterize new devices. New phenomena need to be modeled requiring modifications or complete reformulation of the existing models. In some cases, suitable equations may not be available or it may be difficult to incorporate the new equations in the existing models while preserving other desirable properties like continuity of current over various operating regions. Measured data or data from device analysis programs can be directly used in these situations in the form of a table model.

The drain current is a function of five variables: $v_{gs}$, $v_{ds}$, $v_{bs}$, $L_e$ and $W_e$. Modeling the drain current as a function of all these variables will lead to the use of five dimensional table, which is impractical. Therefore, knowledge of the device behavior is used to reduce this five dimensional table into several tables of reduced dimensionality.

The threshold voltage is principally a function of $v_{bs}$. The second order phenomena affecting the threshold voltage are the short channel effect, the narrow width effect and drain induced barrier lowering. Thus, the threshold voltage is modeled using three tables:

$$vth = VT1(v_{bs}) + VT2(v_{bs}, \frac{1}{W_e}) + VT3(v_{ds}, \frac{1}{L_e})$$

A transformation is then made to a new variable $v_{gst} = v_{gs} - vth$. The primary component of the drain current can then be modeled by one three dimensional table, one two dimensional table or by one two dimensional and one one dimensional table:

$$IDO = ID3(v_{gst}, v_{ds}, v_{bs})$$

or,

$$IDO = ID2(v_{gst}, v_{ds})$$

or,

$$IDO = ID2(v_{gst}, v_{ds})\ ID1(v_{bs})$$

The primary table is normalized by dividing the measured or simulated drain current by $W_e\ /\ L_e$. In the subthreshold region of operation, drain current is an exponential function of terminal voltages and therefore, for $v_{gst} < 0$, it is convenient to store the logarithm of drain current. The drain current is affected by three different electric fields in the device: a vertical electric field, an electric field parallel to the current flow and an electric field perpendicular to the current flow. Hence three correction factors are applied to the primary drain current table:

$$i_{ds} = \frac{W_e}{L_e} \, IDO \times$$

$$IDL\,1(v_{gst}, \frac{1}{L_e}) \; IDL\,2(v_{ds}, \frac{1}{L_e}) \; IDW(v_{bs}, \frac{1}{W_e})$$

Table dimensions can be chosen based on sensitivities of device characteristics. Less number of points are required in the $v_{bs}$ dimension compared to the $v_{gst}$ and $v_{ds}$ dimensions, since the drain current is a less sensitive function of $v_{bs}$. Similarly, less number of points are generally required in the width dimension than the length dimension, since the drain current scales better with the width than the length dimension. Use of the second order correction tables extends the usable range of tables over a larger geometry range.

Some type of interpolation scheme has to be used when the data values are required for intermediate values of the independent variables. Polynomial interpolation can result in nonmonotonic behavior of drain current and also may be difficult to implement in multiple dimensions. More complex schemes are available which assure monotonic behavior [56]. However, use of more complex interpolation schemes increases the model evaluation time, diminishing some of the advantage of table lookup models. Superposition of linear interpolation can be used to obtain higher computation speed, at the same time maintaining the monotonocity of drain current.

Measured or simulated data for various size devices can be easily transformed into the needed tables. At the same time, some of the basic device parameters, such as the threshold voltage and mobility, can be conveniently extracted, giving useful information for process control.

Table lookup models are generally used for the dc characteristics only. They are not used for charge storage because the charges associated with the device terminals are very difficult to measure. It is possible to compute the charges from terminal capacitances [62], but even the capacitances are difficult to measure, especially for small size devices. However, it is possible to obtain terminal charges from device analysis programs. The model computation time for charges and capacitances is quite small for analytical equations and table lookup models do not provide significant advantage in that respect.

Temperature dependence of device characteristics can be taken into account by having two additional tables, one for threshold voltage and one for drain current:

$$vth = vth + VTT(T)$$

$$i_{ds} = i_{ds} \; IDT(T)$$

## 5.24. Depletion devices

All the models described so far in this chapter predict the device characteristics of *normally off* or *enhancement* type devices. These devices are *off* when zero voltage is applied between the gate and source and the device turns *on* and the drain current increases as this voltage is increased. The *depletion* type device is *normally on*, i.e. substantial drain current flows with zero gate to source voltage and the drain current decreases as this voltage is decreased. Depending on the doping levels, this type of device may or may not be completely turned off even with sufficiently negative gate to source bias. The depletion mode device is usually formed by introducing a buried channel i.e. a channel underneath the surface, usually by ion implantation. Surface channel may also exist depending on the bias conditions.

With the advent of the CMOS technology, many circuits are designed using enhancement type p- and n-channel devices. Some circuits use depletion mode MOSFETs to achieve improved performance, enhanced speed, reduced power and smaller die size. They are most commonly used as load elements in enhancement-depletion logic to decrease the power-delay product. Depletion mode devices, however, have also been used extensively in more general configurations to utilize some of their superior qualities such as higher mobility, low surface generated $1/f$ noise, higher breakdown voltage and reduced threshold voltage sensitivity to channel length.

The behavior of the depletion mode device is generally modeled using the enhancement device models with a proper threshold voltage and mobility shift. This technique is adequate if the depletion devices are used in only one particular configuration, for example with gate and source connected together. Such modeling is inadequate for depletion devices operated under varying operating conditions [63]. Operation of the depletion device is much more complex compared to the enhancement device. The depletion device has many more regions of operation and the drain current equations describing the device behavior are computationally complex [64] [65]. A simpler model is obtained by modifying the CSIM or the BSIM models with the introduction of two additional model parameters:

$DELTA$ = threshold voltage shift due to depletion implant

$RHO$ = ratio of surface to bulk mobilities

Two intermediate quantities are defined:

$$v_{transition} = vth$$

$$+ DELTA \left\{ 1 \frac{K1 \sqrt{PHI- v_{bs}} - K2 \left(PHI- v_{bs}\right)}{DELTA} \right\}$$

If $v_{gs} \leq v_{transition}$,

$$v_{drive} = v_{gs} - vth$$

If $v_{gs} > v_{transition}$,

$$v_{drive} = \frac{1}{2} \left( v_{gs} + v_{transition} \right) - vth$$

$$+ \rho \left( v_{gs} - v_{transition} \right)$$

$$\rho = \frac{RHO}{2 \left[ 1 + U_{gs} \left( v_{gs} - v_{transition} \right) \right]}$$

In all the model equations, $v_{gs} - vth$ is replaced by $v_{drive}$ and in the final $i_{ds}$ equation, $\beta_{eff}$ is replaced by $\beta_o$. All the zero field mobility parameters now describe the bulk mobility behavior and *VFB* or *VTO* is to be adjusted so that it gives the effective threshold voltage including the effect of the buried channel implant.

## 5.25. Model parameter extraction

A large number of MOSFET models is described in this chapter. However, a general model parameter extraction procedure is described in this section to keep the discussion short and manageable. MOSFET models are complex and a large amount of data is needed to accurately extract the parameters for these models. Therefore an automated measurement system is generally used to measure device characteristics and an optimization program is used to extract the model parameters.

Different size transistors are used to cover the desired range of device geometries. A long and wide channel transistor can be used as a reference device. Sometimes it is better to choose the most commonly used device as the reference device so that it will be modeled with maximum accuracy. Ideally, in addition to the reference device, one device with a different channel length and one device with a different channel width is needed to extract the geometry dependence. But in practice it is recommended that about

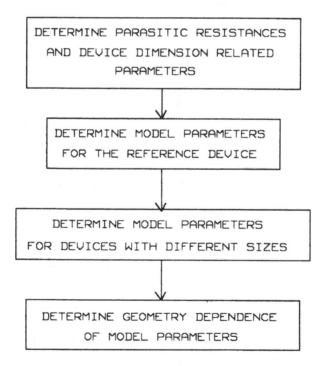

**Figure 5.9:** Parameter extraction procedure for MOSFET models.

three devices with different channel lengths and three devices with different channel widths be used [47]. The overall
model parameter extraction procedure is shown schematically
in Figure 5.9. Initially the parasitic resistances and model
parameters determining the device dimensions are extracted
as explained in an earlier section. The remaining model
parameters for the reference device are then extracted.
After extracting the model parameters for different size devices, parameters related to the length and width dependence
are determined. It is important to choose devices of proper
dimensions so that the parameter has a dominating effect on
the device characteristics to enable an accurate extraction of
its geometry dependence.

Judicious use of an optimization program is made for
extracting model parameters of individual transistors. The
optimizer is used in conjunction with the sequential regionalized approach as shown in Figure 5.10. A family of drain
current curves is measured as a function of $v_{gs}$ for different
values of $v_{bs}$ and a constant $v_{ds}$. $v_{ds}$ is fixed at a small constant voltage, typically 0.1 Volts, to make sure that the device is operating in the linear region. Good initial estimate of
threshold voltage can be obtained by extrapolating the region
of $i_{ds}$ versus $v_{gs}$ curve for fixed $v_{bs}$ having maximum slope.
Threshold voltage can be obtained as a function of $v_{bs}$ by
repeating this procedure for curves with different constant
$v_{bs}$ values. This information is used to obtain initial estimates of threshold voltage related parameters. The maximum slope of the $i_{ds}$ versus $v_{gs}$ curve at zero substrate bias
gives an initial estimate of zero bias or low field mobility.
The $i_{ds}$ values at higher $v_{gs}$ from this curve give estimates of
mobility degradation parameters related to $v_{gs}$. Estimates of
parameters for mobility degradation due to $v_{bs}$ can be
obtained from the variation of the maximum slope for

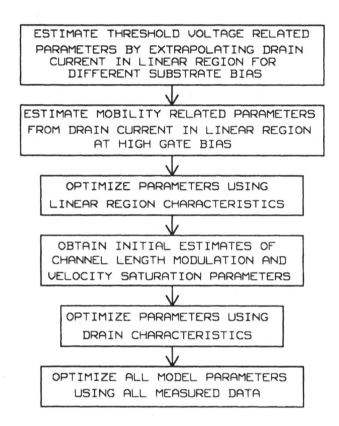

**Figure 5.10:** Parameter extraction procedure for individual MOSFET devices.

different constant $v_{bs}$ curves. After obtaining good initial estimates of all the linear region parameters, the optimizer is then used to refine the model parameter values using all the linear region data. Drain current is then measured as a function of $v_{ds}$ for different constant values of $v_{gs}$ keeping $v_{bs}$ constant. This data is referred as the drain characteristics. Initial estimates are obtained for parameters related to velocity saturation, channel length modulation and saturation region conduction using the drain current data at high $v_{ds}$ to make sure that the device is operating in the saturation region. These parameters are then optimized using the drain characteristics measured for different values of $v_{bs}$. Overall optimization is then performed for all model parameters using all the measured data. It is important to monitor the residual error in the drain current. A consistent pattern may indicate inadequacy of the model to accurately reproduce the device characteristics. The model may be inadequate in some particular regions of device operation. The difference between the final model parameter values and the initial estimates should also be monitored. Large differences may suggest a change in the extraction procedure or change in the bias range of measured data used to extract these particular parameters.

The procedure for extracting the model parameters related to the parasitic diodes was described in an earlier chapter. Special test structures are usually used to extract parameters for the bottom and sidewall (periphery) junctions.

## 5.26. Charge storage

Charge storage in the MOSFET consists of capacitances associated with parasitics and the intrinsic device. The parasitics consist of the overlap capacitances of the gate with respect to the other three terminals and the capacitances of the junction diodes, bottom and sidewall (periphery), of the bulk-source and the bulk-drain junctions. Modeling of the diode capacitances was discussed earlier. Several approaches for modeling the charge storage for the intrinsic device are described here.

## 5.27. Meyer capacitance model

The Meyer charge model computes the MOSFET charges using three voltage dependent capacitances from the gate terminal to each of the other three device terminals

[66], as shown in Figure 5.11.

$$\gamma = \frac{vth - vbi}{sarg}$$

$$\Phi_f = \frac{1}{2} PHI$$

$$C_o = COX \ W_e \ L_e$$

$$v_{gb} = v_{gs} - v_{bs}$$

Cut-off region ($v_{gs} \leq vth$):

$$C_{gs} = 0$$

$$C_{gd} = 0$$

$$C_{gb} = \begin{cases} \dfrac{C_o}{\left[1 + \dfrac{4}{\gamma^2}\left(v_{gb} - vfb\right)\right]^{\frac{1}{2}}} & \text{if } v_{gb} > vfb \\ C_o & \text{if } v_{gb} \leq vfb \end{cases}$$

Saturation region ($v_{ds} \geq v_{dsat}$):

$$C_{gs} = \frac{2}{3} C_o$$

$$C_{gd} = 0$$

$$C_{gb} = 0$$

Linear region ($v_{ds} < v_{dsat}$):

$$C_{gs} = \frac{2}{3} C_o \left[1 - \frac{\left(v_{dsat} - v_{ds}\right)^2}{\left(2 \ v_{dsat} - v_{ds}\right)^2}\right]$$

$$C_{gd} = \frac{2}{3} C_o \left[ 1 - \frac{v_{dsat}{}^2}{\left( 2\, v_{dsat} - v_{ds} \right)^2} \right]$$

$$C_{gb} = 0$$

## 5.28.  Smoothed Meyer model

The capacitance versus voltage characteristics for the Meyer model have several discontinuities. These discontinuities may cause slow convergence of the Newton-Raphson iterations and in some cases convergence may be prevented entirely. One remedy for the problem of discontinuities is to smooth the capacitance characteristics [67], as shown in Figure 5.11.

Cut-off region ($v_{gs} \leq vth$):

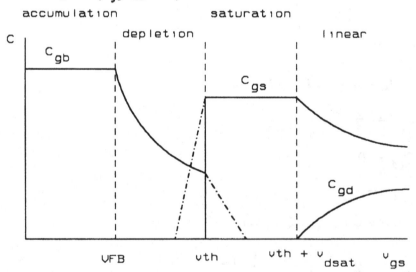

**Figure 5.11:**  Capacitance characteristics of Meyer model and smoothing of discontinuities.

$$C_{GSC} = \frac{2}{3} C_o \frac{v_{gs} - (\ vth - \Phi_f\ )}{\Phi_f}$$

$$C_{gs} = \begin{cases} 0 & \text{if } v_{gs} \leq vth - \Phi_f \\ C_{GSC} & \text{if } vth - \Phi_f < v_{gs}, v_{ds} \geq 0.1 \\ C_{GSC} \left( \frac{0.1 + v_{ds}}{0.2} \right) & \text{if } vth - \Phi_f < v_{gs}, v_{ds} < 0.1 \end{cases}$$

$$C_{gd} = \begin{cases} 0 & \text{if } v_{gs} \leq vth - \Phi_f \\ 0 & \text{if } vth - \Phi_f < v_{gs}, v_{ds} \geq 0.1 \\ C_{GSC} \left( \frac{0.1 - v_{ds}}{0.2} \right) & \text{if } vth - \Phi_f < v_{gs}, v_{ds} < 0.1 \end{cases}$$

$$C_{gb} = \begin{cases} \dfrac{C_o}{\left[ 1 + \dfrac{4}{\gamma^2} (\ v_{gb} - vfb\ ) \right]^{\frac{1}{2}}} & \text{if } v_{gb} > vfb \\ C_o & \text{if } v_{gb} \leq vfb \end{cases}$$

On region $(v_{gs} > vth)$:

$$C_{GBO} = \begin{cases} \dfrac{C_o}{\left[ 1 + \dfrac{4}{\gamma^2} (\ v_{gb} - vfb\ ) \right]^{\frac{1}{2}}} & \text{if } v_{gb} > vfb \\ C_o & \text{if } v_{gb} \leq vfb \end{cases}$$

$$C_{gb} = \begin{cases} C_{GBO} \dfrac{-\ v_{gs} + vth + PHI}{PHI} & v_{gs} < vth + PHI \\ 0 & v_{gs} \geq vth + PHI \end{cases}$$

Peak region ($v_{gs} - vth < 0.1$):

$v_{ds} < 0.1$:

$$C_{GS1} = \frac{2}{3} C_o \frac{0.1 + v_{ds}}{0.2}$$

$$C_{GD1} = \frac{2}{3} C_o \frac{0.1 - v_{ds}}{0.2}$$

$$C_{GS2} = \frac{2}{3} C_o \left[ 1 - \frac{\left( 0.1 - v_{ds} \right)^2}{\left( 0.2 - v_{ds} \right)^2} \right]$$

$$C_{GD2} = \frac{2}{3} C_o \left[ 1 - \frac{0.01}{\left( 0.2 - v_{ds} \right)^2} \right]$$

$$C_{gs} = \left( C_{GS2} - C_{GS1} \right) \frac{v_{gs} - vth}{0.1} + C_{GS1}$$

$$C_{gd} = \left( C_{GD2} - C_{GD1} \right) \frac{v_{gs} - vth}{0.1} + C_{GD1}$$

$v_{ds} \geq 0.1$:

$$CGS = \frac{2}{3} C_o$$

$$CGD = 0$$

Transition region ($v_{gs} - vth \geq 0.1$ , $v_{ds} < 0.1$):

$$C_{gs} = \frac{2}{3} C_o \left[ 1 - \frac{\left( 0.1 - v_{ds} \right)^2}{\left( 0.2 - v_{ds} \right)^2} \right]$$

$$C_{gd} = \frac{2}{3} C_o \left[ 1 - \frac{0.01}{\left( 0.2 - v_{ds} \right)^2} \right]$$

Saturation region ( $v_{gs} - vth \geq 0.1$ , $v_{ds} \geq v_{dsat}$ ):

$$C_{gs} = \frac{2}{3} C_o$$

$$C_{gd} = 0$$

Linear region ( $v_{gs} - vth \geq 0.1$ , $v_{ds} < v_{dsat}$ ):

$$C_{gs} = \frac{2}{3} C_o \left[ 1 - \frac{\left( v_{dsat} - v_{ds} \right)^2}{\left( 2 v_{dsat} - v_{ds} \right)^2} \right]$$

$$C_{gd} = \frac{2}{3} C_o \left[ 1 - \frac{v_{dsat}^2}{\left( 2 v_{dsat} - v_{ds} \right)^2} \right]$$

Smoothing of the capacitance characteristics has several drawbacks. Firstly, smoothing introduces errors in the capacitances. The amount of error is proportional to the effectiveness of the smoothing alleviating the numerical difficulties. Secondly, care must be taken in performing the smoothing since the capacitances are functions of more than one voltage. Smoothing must therefore be done in multiple dimensions. Thirdly, smoothing destroys the relationship between the capacitances and the charges dictated by device physics. The capacitances no longer represent the derivatives of the charges.

## 5.29. Charge nonconservation problem

The problem associated with modeling charge storage using capacitance equations is explained below [6] [68] [69].

The current through a nonlinear voltage dependent capacitor is given by,

$$i = \frac{\partial q}{\partial t} \qquad (5.43)$$

If the capacitance equation is available, then the above equation is usually approximated by,

$$i \approx C(v) \frac{\partial v}{\partial t} \qquad (5.44)$$

In the circuit simulation program, this equation is integrated to give,

$$\int_{t_n}^{t_{n+1}} i \, dt = \hat{C} \left( v_{n+1} - v_n \right) \qquad (5.45)$$

where $\hat{C}$ represents the capacitance over the voltage interval between $v_n$ and $v_{n+1}$. $\hat{C}$ is approximated by several commonly used choices:

$$\hat{C} = \begin{cases} C_{n+1} \\ C_n \\ \frac{1}{2} \left( C_n + C_{n+1} \right) \end{cases} \qquad (5.46)$$

None of the above choices gives an exact integral of capacitance. The exact relationship is given by,

$$\int_{t_n}^{t_{n+1}} i \, dt = q_{n+1} - q_n \qquad (5.47)$$

Therefore we need to compute $q(v)$. Since $C(v)$ is available, this is approximated by,

$$q \approx C(v) \; v \qquad (5.48)$$

as done in equation (5.44) and,

$$\frac{\partial q}{\partial t} \approx C(v)\, \frac{\partial v}{\partial t} \tag{5.49}$$

which is then integrated using numerical techniques. This amounts to neglecting the second term in the exact relationship:

$$\frac{\partial q}{\partial t} = C(v)\, \frac{\partial v}{\partial t} + v\, \frac{\partial C}{\partial t} \tag{5.50}$$

The above two approximations destroy the exact relationship between current and stored charge and hence voltage. Thus while the integral of current between times $t_1$ and $t_2$ may be zero, the charges and voltages at the two times may differ due to the errors introduced by the approximations. That is,

$$\int_{t_1}^{t_2} i\; dt = 0 \; but \; q(t_2) \neq q(t_1) \; and \; v(t_2) \neq v(t_1)$$

Alternatively, the charges and voltages at times $t_1$ and $t_2$ may be the same, while the integral of current is nonzero.

$$q(t_2) = q(t_1) \; and \; v(t_2) = v(t_1) \; but \int_{t_1}^{t_2} i\; dt \neq 0$$

Stated another way, $q$ is approximated by,

$$q = \int C\; dv \tag{5.51}$$

and the integral is computed numerically. Therefore, the integral will depend on the path of the voltage variable. Starting at the same initial voltage, if the final voltage is reached by following two different paths, the resulting charges may be different. Therefore, it is necessary to have analytical equations for $q$ instead of computing $q$ using numerical integration from the analytical equations for $C$.

Although illustrated for the case of a two terminal capacitor, nonconservation of charge can occur in any device in which the capacitance is a function of the terminal voltage. The magnitude of the error is proportional to the degree of nonlinearity. The problem becomes even worse for multiterminal devices where the terminal capacitances are not only the functions of the voltage across the capacitors but of other terminal voltages as well. Thus, in general, for multiterminal devices, the charges associated with each terminal are functions of all the terminal voltages. The derivatives of the charges, therefore, are represented by a capacitance matrix where the capacitances are given by,

$$C_{km} = \frac{\partial q_k}{\partial v_m} \qquad (5.52)$$

It is to be noted that the capacitances, in general, are nonreciprocal, i.e.,

$$C_{km} \neq C_{mk} \qquad (5.53)$$

The charge nonconservation is illustrated by a simple circuit shown in Figure 5.12, where the MOS transistor is used as a voltage dependent capacitor in series with a linear capacitor, an arrangement commonly used in bootstrap circuits. After a number of pulses, the voltage on the capacitor will be proportional to the charge transferred through the gate oxide of the MOS transistor. For a practical device with no gate leakage there will be no charge transferred. As shown in the simulated waveforms using the capacitance model, however, either positive or negative charge is tranferred depending on the approximation used for $\hat{C}$. The charge based model produces the correct waveform.

### 5.30.  Charge based models

The charge based method assumes that the current in each terminal is given as the derivative of a stored charge. While this assumption holds for simple capacitors, complications arise when the charge based method is applied to more complex devices such as the MOS transistor. The MOS transistor conducts current under steady state bias

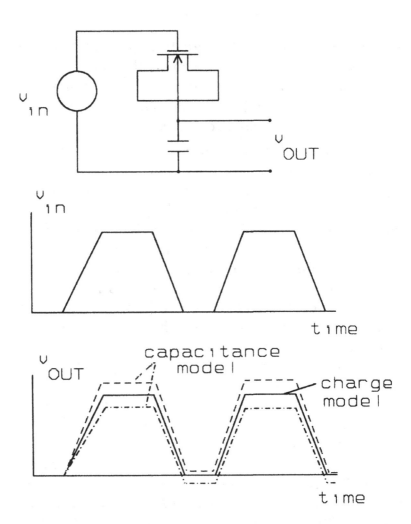

**Figure 5.12:** The charge nonconservation problem.

conditions. The dc component of current does not correspond to a change in stored charge and thus cannot be modeled by the charge based method. The device is therefore modeled by superposition of a charge model and a dc model. The current in each terminal is:

$$i(t) = i_{dc}(v) + i_{tran}(t) \tag{5.54}$$

where the dc current $i_{dc}$ depends only on the instantaneous terminal voltages and the transient component $i_{tran}$ is zero under steady state conditions. $i_{dc}$ is represented by an appropriate dc model while the charge based model is used to compute the transient current $i_{tran}$.

For the MOS transistor, gate, bulk, source and drain charge equations are needed. Exact calculation of charges $Q_g$, $Q_b$, $Q_s$ and $Q_d$ as function of time is difficult, requiring solution of nonlinear partial differential device equations. A solution which is valid at relatively low circuit speeds, i.e. speeds lower than the carrier transit time through the device, may be obtained by invoking the quasi-static approximation. The transistor charges are calculated as functions of terminal voltages under steady state conditions. It is then assumed that the resulting relationship holds even during transients. Thus the charging currents in the channel are ignored and the MOS transistor is treated as a simple multiterminal capacitance. A uniform channel approximation is made to simplify the Poisson's equation. The gradual channel approximation states that the electric field parallel to the channel is small compared to the field perpendicular to the channel and allows Poisson's equation to be reduced to one dimension. The uniform channel approximation further assumes that the device is uniform along the length of the channel so that the inversion layer charge $Q_n(y,t)$ may be written in terms of the channel potential $v(y,t)$ and the gate voltage $v_g$:

$$Q_n(y,t) = Q_n\bigl(v(y,t)\,,\,v_g(t)\bigr) = Q_n(v,v_y) \tag{5.55}$$

and,

$$\mu(y,t) = \mu\left(v(y,t)\,,\,v_g(t)\right) = \mu(v,v_y) \qquad \textbf{(5.56)}$$

where $y$ is the dimension along the length of the channel.

### 5.31. Ward-Dutton charge model

This model computes the gate and bulk charges from one dimensional Poisson equation applied perpendicular to the surface [68] [69]. Channel charge is computed from:

$$Q_c = -\left(Q_g + Q_b\right)$$

and is related to the source and drain charges,

$$Q_c = Q_s + Q_d$$

Unfortunately, the drain and source charges are not as well defined as the gate and bulk charges and must be obtained by partitioning the channel charge. The source and drain charges can be expressed as:

$$Q_s = -W_e \int_0^{L_e} \left(1 - \frac{y}{L_e}\right) Q_n\, dy$$

$$Q_d = -W_e \int_0^{L_e} \frac{y}{L_e}\, Q_n\, dy$$

but the integration is tedious and too complicated for use in circuit simulation allpications. SPICE uses the following scheme for partitioning the channel charge:

$$Q_d = \begin{cases} \dfrac{1}{2}\,Q_c & \text{\textit{linear region}} \\ XQC\,Q_c & \text{\textit{saturation region}} \end{cases}$$

$$Q_s = \begin{cases} \dfrac{1}{2} Q_c & \textit{linear region} \\ (1 - XQC) Q_c & \textit{saturation region} \end{cases}$$

where $XQC \leq 0.5$ and is a user input model parameter. This partitioning scheme causes a discontinuity at the boundary of the linear and saturation regions. The charges will be continuous if $XQC = \dfrac{1}{2}$, but then $Q_d$ is nonzero in saturation region which is reported to result in excessive gate-to-drain feedthrough contrary to measured circuit performance. It is possible to use some smoothing function but its use in practice still causes some nonconvergence problems.

$$\gamma = \frac{vth - vbi}{sarg}$$

$$C_o = COX \; W_e \; L_e$$

$$v_g = v_{gs} - v_{bs} - vbi + PHI$$

$$v_s = PHI - v_{bs}$$

$$v_z = \begin{cases} PHI - v_{bs} + v_{ds} & v_{ds} \leq v_{dsat} \\ PHI - v_{bs} + v_{dsat} & v_{ds} > v_{dsat} \end{cases}$$

Accumulation region ($v_g \leq 0$):

$$Q_g = C_o \; v_g$$

$$Q_b = - \; Q_g$$

$$Q_c = 0$$

$$Q_d = 0$$

$$Q_s = 0$$

Cut-off region ($v_{gs} \leq vth$):

$$Q_g = C_o \, \gamma \left[ \sqrt{\frac{\gamma^2}{4} + v_g} - \frac{\gamma}{2} \right]$$

$$Q_b = - \, Q_g$$

$$Q_c = 0$$

$$Q_d = 0$$

$$Q_s = 0$$

On region ($v_{gs} > vth$):

$$i = v_g \left( v_z^{\frac{1}{2}} + v_s^{\frac{1}{2}} \right) - \frac{2}{3} \, \gamma \left( v_z + v_z^{\frac{1}{2}} v_s^{\frac{1}{2}} + v_s \right)$$

$$- \frac{1}{2} \left( v_z + v_s \right) \left( v_z^{\frac{1}{2}} + v_s^{\frac{1}{2}} \right)$$

$$Q_g = C_o \, v_g - \frac{C_o}{i} \left\{ \frac{1}{2} \, v_g \left( v_z + v_s \right) \left( v_z^{\frac{1}{2}} + v_s^{\frac{1}{2}} \right) \right.$$

$$- \frac{2}{5} \, \gamma \left[ v_z^2 + v_z^{\frac{1}{2}} v_s^{\frac{1}{2}} \left( v_z + v_s \right) + v_s^2 \right]$$

$$\left. - \frac{1}{3} \left( v_z^{\frac{1}{2}} + v_s^{\frac{1}{2}} \right) \left( v_z^2 + v_z \, v_s + v_s^2 \right) \right\}$$

$$Q_b = - \frac{C_o \gamma}{i} \left\{ \frac{2}{3} v_g \left( v_z + v_z^{\frac{1}{2}} v_s^{\frac{1}{2}} + v_s \right) \right.$$

$$- \frac{1}{2} \gamma \left( v_z + v_s \right) \left( v_z^{\frac{1}{2}} + v_s^{\frac{1}{2}} \right)$$

$$\left. - \frac{2}{5} \left[ v_z^2 + v_z^{\frac{1}{2}} v_s^{\frac{1}{2}} \left( v_z + v_s \right) + v_z v_s + v_s^2 \right] \right\}$$

$$Q_c = - \left( Q_g + Q_b \right)$$

$$Q_d = \begin{cases} \frac{1}{2} Q_c & v_{ds} \leq v_{dsat} \\ XQC \ Q_c & v_{ds} > v_{dsat} \end{cases}$$

$$Q_s = \begin{cases} \frac{1}{2} Q_c & v_{ds} \leq v_{dsat} \\ (1 - XQC) \ Q_c & v_{ds} > v_{dsat} \end{cases}$$

## 5.32. Yang-Chatterjee charge model

This model pays special attention to make sure that the charges and the capacitances are continuous throughout different regions [6]. It is assumed that in the saturation region, the channel is isolated from the drain and all the channel mobile charge $Q_c$ is associated with $Q_s$ and $Q_d$ is zero. In the saturation region, the charges electrostatically associated with the drain are the mobile charges beyond the pinchoff point and the drain depletion charges. Since these two charges are neglected in the computation of $Q_c$, the partitioning of all the channel charge $Q_c$ into $Q_s$ is a reasonable assumption. Since $Q_d$ is zero in saturation, the excessive gate-to-drain feedthrough is not observed, giving more *Meyer like* behavior.

$$\gamma = \frac{vth - vbi}{sarg}$$

$$\alpha_x = \frac{v_{gs} - vth}{v_{dsat}}$$

$$C_o = COX \ W_e \ L_e$$

Accumulation region ($v_{gs} \leq vfb + v_{bs}$):

$$Q_g = C_o \left( v_{gs} - vfb - v_{bs} \right)$$

$$Q_b = - \ Q_g$$

$$Q_c = 0$$

$$Q_s = 0$$

$$Q_d = 0$$

Cut-off region ($vfb + v_{bs} < v_{gs} \leq vth$):

$$Q_g = C_o \frac{\gamma^2}{2} \left\{ - 1 + \left[ 1 + \frac{4 \left( v_{gs} - vfb - v_{bs} \right)}{\gamma^2} \right]^{\frac{1}{2}} \right\}$$

$$Q_b = - \ Q_g$$

$$Q_c = 0$$

$$Q_s = 0$$

$$Q_d = 0$$

Saturation region ($vth < v_{gs} \leq \alpha_x \, v_{ds} + vth$):

$$Q_g = C_o \left[ v_{gs} - vfb - PHI - \frac{( \, v_{gs} - vth \, )}{3 \, \alpha_x} \right]$$

$$Q_b = C_o \left[ vfb + PHI - vth \right.$$
$$\left. \frac{( 1 - \alpha_x ) ( \, v_{gs} - vth \, )}{3 \, \alpha_x} \right]$$

$$Q_c = - \frac{2}{3} \, C_o \, ( \, v_{gs} - vth \, )$$

$$Q_d = 0$$

$$Q_s = - \frac{2}{3} \, C_o \, ( \, v_{gs} - vth \, )$$

Linear region ($v_{gs} > \alpha_x \, v_{ds} + vth$):

$$Q_g = C_o \left[ v_{gs} - vfb - PHI - \frac{v_{ds}}{2} \right.$$
$$\left. + \frac{\alpha_x \, v_{ds}^2}{12 \left( v_{gs} - vth - \dfrac{\alpha_x \, v_{ds}}{2} \right)} \right]$$

$$Q_b = C_o \left[ vfb + PHI - vth + \frac{1 - \alpha_x}{2} v_{ds} \right.$$

$$\left. - \frac{\left( 1 - \alpha_x \right) \quad \alpha_x \, v_{ds}^2}{12 \left( v_{gs} - vth - \dfrac{\alpha_x \, v_{ds}}{2} \right)} \right]$$

$$Q_c = - C_o \left[ v_{gs} - vth - \frac{\alpha_x}{2} v_{ds} \right.$$

$$\left. + \frac{\alpha_x^{\,2} \, v_{ds}^2}{12 \left( v_{gs} - vth - \dfrac{\alpha_x \, v_{ds}}{2} \right)} \right]$$

$$Q_d = - C_o \left[ \frac{\left( v_{gs} - vth \right)}{2} - \frac{3}{4} \alpha_x \, v_{ds} \right.$$

$$\left. + \frac{\alpha_x^{\,2} \, v_{ds}^2}{8 \left( v_{gs} - vth - \dfrac{\alpha_x \, v_{ds}}{2} \right)} \right]$$

$$Q_s = - C_o \left[ \frac{\left( v_{gs} - vth \right)}{2} - \frac{1}{4} \alpha_x \, v_{ds} \right.$$

$$\left. - \frac{\alpha_x^{\,2} \, v_{ds}^2}{24 \left( v_{gs} - vth - \dfrac{\alpha_x \, v_{ds}}{2} \right)} \right]$$

### 5.33. BSIM charge model

This model gives significant attention to the partitioning of the channel charge [51] [52]. The proposed partitioning changes smoothly from 40/60 for $Q_d/Q_s$ in the saturation region asymptotically to 50/50 in the linear region. But the nonzero drain charge in saturation causes problems similar to those observed for the Ward-Dutton model. Therefore, this model also offers an alternative partition with zero $Q_d$ in the saturation region. Under this condition, the model reduces to the Yang-Chatterjee model described above.

The equations for $Q_s$ are not given since it can be computed from:

$$Q_g + Q_b + Q_d + Q_s = 0$$

$$\gamma = \frac{vth - VFB - PHI}{sarg}$$

$$\alpha_x = \frac{v_{gs} - vth}{v_{dsat}}$$

$$C_o = COX \; W_e \; L_e$$

Accumulation region ($v_{gs} \leq VFB + v_{bs}$):

$$Q_g = C_o \left( v_{gs} - VFB - v_{bs} \right)$$

$$Q_b =- Q_g$$

$$Q_d = 0$$

Cut-off region ($VFB + v_{bs} < v_{gs} \leq vth$):

$$Q_g = C_o \frac{\gamma^2}{2} \left\{ -1 \right.$$

$$\left. + \left[ 1 + \frac{4 \left( v_{gs} - VFB - v_{bs} \right)}{\gamma^2} \right]^{\frac{1}{2}} \right\}$$

$$Q_b = - Q_g$$

$$Q_d = 0$$

Saturation region ($vth < v_{gs} \leq \alpha_x v_{ds} + vth$):

$$Q_g = C_o \left[ v_{gs} - VFB - PHI - \frac{\left( v_{gs} - vth \right)}{3 \alpha_x} \right]$$

$$Q_b = C_o \left[ VFB + PHI - vth \right.$$

$$\left. - \frac{\left( 1 - \alpha_x \right) \left( v_{gs} - vth \right)}{3 \alpha_x} \right]$$

$$Q_d = \begin{cases} 0 & XPART = 1 \\ -\dfrac{4}{15} C_o \left( v_{gs} - vth \right) & XPART = 0 \end{cases}$$

Linear region ($v_{gs} > \alpha_x v_{ds} + vth$):

$$Q_g = C_o \left[ v_{gs} - VFB - PHI - \frac{v_{ds}}{2} \right.$$

$$\left. + \frac{\alpha_x\, v_{ds}^2}{12\left( v_{gs} - vth - \dfrac{\alpha_x\, v_{ds}}{2} \right)} \right]$$

$$Q_b = C_o \left[ VFB + PHI - vth + \frac{1 - \alpha_x}{2}\, v_{ds} \right.$$

$$\left. - \frac{\left( 1 - \alpha_x \right) \quad \alpha_x\, v_{ds}^2}{12\left( v_{gs} - vth - \dfrac{\alpha_x\, v_{ds}}{2} \right)} \right]$$

If $XPART = 1$,

$$Q_d = - C_o \left[ \frac{\left( v_{gs} - vth \right)}{2} - \frac{3}{4}\, \alpha_x\, v_{ds} \right.$$

$$\left. + \frac{\alpha_x{}^2\, v_{ds}^2}{8\left( v_{gs} - vth - \dfrac{\alpha_x\, v_{ds}}{2} \right)} \right]$$

If $XPART = 0$,

$$Q_d = - C_o \left\{ \frac{vgs - vth}{2} - \frac{\alpha_x \, vds}{2} \right.$$

$$+ \frac{\alpha_x \, v_{ds}}{\left( v_{gs} - vth - \frac{\alpha_x \, v_{ds}}{2} \right)^2}$$

$$\left. \left[ \frac{\left( v_{gs} - vth \right)^2}{6} - \frac{\alpha_x \, v_{ds} \left( v_{gs} - vth \right)}{8} + \frac{\alpha_x^{\,2} \, v_{ds}^2}{40} \right] \right\}$$

## 5.34. Second order effects on charges

The uniform channel approximation assumes that the source and drain potentials affect the channel charge only through the channel voltage $v(y,t)$. While this assumption is valid over most of the channel length for long devices, it breaks down for short channel transistors [69]. Furthermore the requirement imposed on the mobility does not take into account other second order effects such as mobility degradation, velocity saturation and channel length modulation. Also, even though the quasi-static approximation makes the charge calculation tractable, it fails at high speeds. Qualitatively, the breakdown of the quasi-static approximation can be explained as follows. Using that assumption, the voltage and charge distributions in a device are determined from steady state calculations. When the terminal voltages vary, these distributions follow, always taking the values that they would have in the steady state. The changes in charge give rise to transient currents. The presence of these transient currents, however, implies a deviation of the voltage and charge from their steady state distributions. For low circuit speeds the transient current is small and these deviations

may be ignored. For high speeds, however, the departure of
the voltage and current in the channel from their steady
state values is large. For long channel devices, where velocity saturation is not observed, the transit time in the transistor is given by,

$$\tau_0 = \frac{L_e^2}{\mu\ v_{dsat}} \qquad (5.57)$$

For short channel devices, the transit time can be approximated by,

$$\tau_0 \approx \frac{L_e}{v_{sat}} \qquad (5.58)$$

We can set a conservative requirement that the rise or fall
times of the input step waveforms should be $\geq 20\ \tau_0$ for the
quasi-static approximation to be valid. For $v_{sat} \approx 50\ km/sec$
and $L_e \approx 1\ \mu m$, $\tau_0 \approx 20\ psecs$. Thus the impact of the
quasi-static approximation on the simulation of most of the
practical circuits is expected to be minor. An implementation of a non-quasi-static model is suggested in [70].
Another way to improve upon the quasi-static approximation
is to divide each device into $n$ lumps, where each lump is
represented by a quasi-static model [69].

All the charge models described here are valid for long
channel devices. Several differences have been observed
between the long channel and the short channel capacitances
[71] - [74], but unfortunately a model which includes the
second order effects is not yet available. $C_{gs}$ and $C_{gd}$ are
much smoother for the short channel device than the long
channel device. Sharp transition between the linear and
saturation regions is not observed. This is due to the velocity saturation effect. $C_{gs}$ curves split in the saturation
region for different $v_{gs}$. Because of the velocity saturation
effect, more channel charge exists under the control of the

gate. This increase is more significant at large gate bias since the velocity saturation effect is more dominant. The capacitance $C_{gd}$ saturates to a finite value in the saturation region. This is due to the channel side fringing fields between the gate and the drain/source junctions. The magnitude of this fringing field capacitance is a strong function of the drain bias. The limiting saturated value of $C_{gd}$ at large $v_{ds}$ is a function of drain junction depth and gate oxide thickness and does not scale with transistor sizes. Therefore, it plays a major role in the shorter transistors.

### 5.35. Temperature dependence

A lot of work needs to be done to determine the functional relationships of the MOSFET model parameters with the operating temperature. Many model parameters are empirical and hence it is not possible to derive their temperature dependence based on device physics. One approach is to use empirical relationships based on practical experience or an alternative approach is to use a different set of model parameters for the desired temperatures of interest: typically, the minimum, the nominal and the maximum temperatures.

The basic common model parameters of mobility and threshold voltage can be computed based on their theoretical temperature dependence. Let,

$TNOM$ = nominal temperature

$T$ = analysis temperature

The mobility is treated as a power law relationship:

$$UO = UO_{TNOM} \left( \frac{T}{TNOM} \right)^{BEX} \tag{5.59}$$

$$KP = KP_{TNOM} \left( \frac{T}{TNOM} \right)^{BEX} \tag{5.60}$$

where BEX is a user input model parameter, typically $-\dfrac{3}{2}$.

The threshold voltage can be computed as:

$$E_g = 1.16 - \frac{7.02 \times 10^{-4} \; T^2}{T + 1108.0} \tag{5.61}$$

$$n_i \; \alpha \; T^{1.5} \; e^{-\frac{E_g}{2kT}} \tag{5.62}$$

$E_{g\,TNOM} = $ band gap at temperature $TNOM$

$E_g = $ band gap at temperature $T$

$$PHI = 2 \; \frac{kT}{q} \ln \frac{NSUB}{n_i} \tag{5.63}$$

$$PHI = 2 \; \frac{T}{TNOM} \; PHI_{TNOM}$$
$$- 2 \; \frac{kT}{q} \left[ \frac{3}{2} \ln \frac{T}{TNOM} + \frac{1}{2k} \left( -\frac{E_g}{T} + \frac{E_{g\,NOM}}{TNOM} \right) \right] \tag{5.64}$$

$$vbi = vbi_{TNOM} - 0.5 \left[ E_g - E_{g\,TNOM} \right] TPG$$
$$+ 0.5 \left[ PHI - PHI_{TNOM} \right] \tag{5.65}$$

$$vfb = vbi - PHI \tag{5.66}$$

$$VTO = vbi + GAMMA \sqrt{PHI} \tag{5.67}$$

Instead of using the above theoretical relationship, sometimes the following empirical equation is used:

$$VTO = VTO_{NOM} \left[ 1 - TVTO \left( T - TNOM \right) \right]$$

where *TVTO* is the first order temperature coefficient.

Most of the remaining model parameters are empirical and generally are not computed as a function of temperature for lack of availability of functional relationships.

The model parameters related to the parasitic diodes are treated similar to the diode model parameters. An empirical relationship is usually used to compute the parasitic resistances:

$$RS = RS_{NOM} \left[ 1 + TRS \left( T - TNOM \right) \right]$$
$$RD = RD_{NOM} \left[ 1 + TRD \left( T - TNOM \right) \right]$$

where *TRS* and *TRD* are first order temperature coefficients.

## 5.36. Modeling for Analog Applications

The MOSFET is no longer a device mainly used for digital applications. Many MOS chips are now designed combining digital and analog functions. Most of the available models are suitable for digital designs. Hence, analog circuits are often designed very conservatively and the full potential of the technology is not realized.

The small signal drain conductance ($g_{ds}$) in saturation is one of the most poorly modeled parameters. One of the reasons contributing to this is that most of the automated parameter extraction programs minimize the error in drain current. But it is possible to get large errors in conductance when the error in drain current is small. The conductance is more important in determining analog circuit performance. One solution is to optimize the model parameters to minimize the error in conductance as well.

Another problem in modeling for analog applications is the failure of the quasi-static approximation. The frequency dependencies of many of the admittances predicted by the quasi-static model do not agree with the expected dependencies as well as with dependencies obtained from detailed numerical simulations. We can say that the quasi-static approximation is valid for frequencies less than $\omega_o$ where $\omega_o$ is of the order of $\dfrac{\mu\ v_{dsat}}{L_e^2}$ or $\dfrac{g_m}{C_{gs}}$. It is possible that this limit may be reached for analog circuits while for digital circuits the quasi-static approximation may still be valid.

Most of the MOS models divide the device operation in several regions and use different equations for each of these regions. This is adequate for digital applications since the devices usually operate in one of these regions and the equations model the device behavior accurately. But in analog applications, the devices may operate at or near the boundaries between these regions and the model equations do not reproduce the device characteristics with acceptable accuracy. For example, there is a region of *moderate* inversion [75] [76] in which neither the *strong* inversion nor the *weak* inversion approximation is valid. Single expressions valid in all regions of operation have been developed considering both drift and diffusion current components. Such expressions avoid artificial transitions from one region to the other and reduce the corresponding large errors in calculating small signal parameters. The mathematical complexity of the very general formulation can be reduced by assuming a *charge sheet* model [77] [78]. The resulting expressions give the drain current in terms of surface potentials rather than terminal voltages. Accurate explicit general equations are not available for the surface potential as a function of terminal voltages. Therefore, implicit equations must be solved numerically, making these models unsuitable for circuit simulation applications. It should also be noted that although these models predict *natural* transitions between various operating regions, they do not take into account other second order effects, needing further refinements.

The small signal conductances and capacitances are obtained by differentiating the large signal drain current and terminal charge equations. These equations are designed to obtain efficient dc and transient simulations. Therefore, another way to obtain accurate ac simulations is to use different models for small signal admittances which will produce accurate results [76]. Of course this approach cannot predict the *distortion* behavior of the circuit since that is determined by the large signal device characteristics.

### 5.37. Power MOSFETs

All the MOSFET models described in this chapter are designed for low power MOSFETs and do not accurately simulate all the modes of power MOSFET operation. Since the power MOSFET is structurally different from its low power counterpart, several aspects of power MOSFET operation are not included in the low power models [79] [80]:

1. off leakage current is not negligible,
2. ohmic gate resistance is not negligible,
3. gate to drain capacitance is voltage dependent,
4. parasitic source resistance is nonlinear,
5. drain to source ohmic resistance in the forward region is not equal to that in the inverse region, since the current paths differ.

Also, the power MOSFET is really a three terminal device with bulk always connected to the source.

The equivalent circuit of a power MOSFET is shown in Figure 5.13. $i_{ds}$ represents the drain current of the intrinsic MOSFET. $C_{gs}$ and $C_{gd}$ are the parasitic capacitances and $C_{gd}$ is usually voltage dependent. $R_g$, $R_d$ and $R_s$ are parasitic resistances. $R_s$ is usually nonlinear and a function of

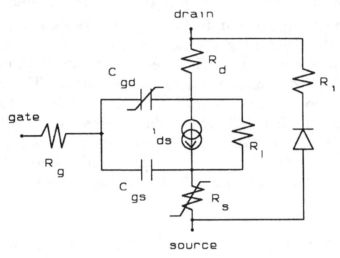

**Figure 5.13:** Power MOSFET equivalent circuit.

drain current. $R_l$ models the leakage current when the device is *off*. The resistance $R_i$ in series with the diode models the current path in the inverse mode of operation, when $v_{ds} < 0$.

Since available MOSFET models are not suitable for power MOSFETs, the approach generally taken is to use an available model and then add external components to improve accuracy. Thus, $i_{ds}$, $C_{gs}$, $C_{gd}$, $R_d$ and $R_s$ are treated as part of the available MOSFET model. Usually, the simple basic level 1 model is used for $i_{ds}$. $R_g$, $R_l$, $R_i$ and the diode are added externally. Thus, $C_{gd}$ and $R_s$ are modeled as constants which degrades the model accuracy. It is possible to model the voltage dependence of $C_{gd}$ by adding a diode between the gate and the drain terminals. The dc current through the diode can be made negligibly small by specifying proper model parameter values. Nonlinear $R_s$ can be modeled by adding a current controlled voltage source in series with the source terminal, where the controlling current is proportional to $i_{ds}$. Simulation inaccuracies may still result due to the fact that the intrinsic MOSFET is treated by the model as a symmetric device. Significant errors may result in simulating switching transients and ringing at the drain terminal where the drain terminal becomes negative with respect to the source terminal.

A more efficient and accurate model of the power MOSFET can be obtained by implementing a different *type* of *element.* Suitable equations can then be used to model all the device characteristics accurately.

# CHAPTER 6

# MESFET MODELS

## 6.1. Introduction

The operation of a metal-semiconductor field-effect transistor (MESFET) is identical to that of a JFET. In the MESFET, however, a metal-semiconductor rectifying contact is used for the gate electrode instead of a p-n junction. The MESFET offers certain processing and performance advantages, such as low temperature formation of the metal-semiconductor barrier (as opposed to a p-n junction made by diffusion or grown processes), low resistance and low $IR$ drop along the channel width, and good heat dissipation for power devices (the rectifying contact can also serve as an efficient thermal contact to heat sink).

Although it is possible to fabricate MESFETs in silicon, gallium arsenide (GaAs) is more commonly used. GaAs MESFETs were used mainly as microwave transistors. But currently GaAs integrated circuits are being developed for a wide variety of applications, ranging from very high volume consumer equipment such as direct broadcast satellite receivers to high performance, low volume items such as ultra high speed digital signal processors. These circuits are typically categorized as either digital or analog. Digital circuits include both logic and memory circuits, while analog circuits include baseband devices such as comparators and analog-to-digital converters as well as the more exotic monolithic microwave integrated circuits [81].

Because of its energy band structure, GaAs is an ideal meduim for achieving remarkable speed in electronic devices and circuits. Electrons in GaAs are exceedingly *light* and highly mobile. The effective mass of electrons is only 7 per cent that of electrons in silicon, and electron mobility in the channels of GaAs MESFETs is about an order of magnitude higher than in silicon MOSFETs. Electron velocities measured in GaAs transistor structures range up to $5 \times 10^7$ meters per second - about five times those achieved in silicon devices. Furthermore, GaAs is readily available in a semi-insulating substrate form that substantially reduces parasitic capacitances, so its potential device speeds can be fully realized in integrated circuits. This high speed, plus power dissipation that tends to substantially lower than that of high speed silicon devices, accounts for the growing interest in GaAs.

The MESFET equivalent circuit is shown in Figure 6.1. Since there are no *standard* MESFET models, a variety of models is being implemented in different versions of SPICE and other circuit simulation programs [82] - [90]. Therefore, a brief review of various proposed model equations is given here. Due to the lack of availability of a MESFET element in many previously used versions of SPICE, another approach is sometimes used to model the MESFETs. A combination of available elements such as JFETs, diodes, resistors and capacitors is used to model the various elements represented in the equivalent circuit of Figure 6.1. The resulting circuit, however, may be more complex and less accurate than the direct implementation of the MESFET equations.

Only one set of model equations is generally used for both *depletion* and *enhancement* type devices. The depletion device is normally *on* allowing flow of substantial drain current. As the gate-to-source voltage is made more negative, for n-channel devices, the width of the depletion region under the metal gate increases, reducing the width of the channel through which current can flow. When gate-to-source voltage is increased beyond pinch-off, the current flow essentially stops. GaAs MESFET integrated circuits give excellent performance [81], but they typically require two power supplies for proper operation and they also must

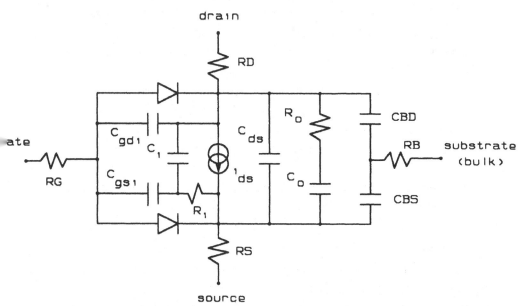

**Figure 6.1:** MESFET equivalent circuit.

contain some voltage level shifting built into the logic gates
to generate the negative gate voltages required for switching
from positive drain voltages. Use of the enhancement device avoids the requirement for dual power supply and level
shifting circuitry by having slightly positive pinch-off voltages. Its structure is similar to the depletion device, except
that the implanted channel depth and the doping concentration are designed so that the built-in potential of the metal
Schottky barrier gate normally cuts off the channel conduction. Thus, a small positive gate voltage (about 0.1 V) must
be applied for source-drain conduction to begin. Enhancement MESFETs are restricted to very limited logic voltage
swings (typically about 0.5 V) because they begin to draw
excessive gate currents for larger gate voltages due to forward biased gate-source and gate-drain diodes. Because the
difference between voltages representing logical 0 and 1
states must be roughly 20 times the standard deviation of
pinch-off (threshold) voltages to allow adequate margins for
building integrated circuits, the pinch-off voltage must be
uniform within 25 mV. It is difficult to achieve this degree
of control. An aditional complication in the manufacture of
enhancement MESFETs is their need for more complex device proecessing than depletion MESFETs. The problem is
that the substantial depletion region at the surface between
gate and the source contact region tends to pinch off the
lightly doped channel. This leads to high source resistance
that degrades the transconductance. This can be avoided by
modifying the fabrication process.

## 6.2. Device symmetry

All the models described in this chapter assume that
$v_{ds} \geq 0$, i.e. the drain is at a higher potential than the
source. If this is not the case, then the roles of the source
and the drain are reversed. Thus, for model computations,
among the topological source and drain terminals, the

terminal which is at a lower potential is to be treated as the
source. The topological connectivity is used to compute the
terminal voltages and also to load the currents and the
charges in the right hand side vector and the jacobians into
the admittance matrix. The topological source and drain
may become electrical darin and source if the device is in
inverse mode of operation. The mapping between the topological and the electrical connections must be done properly
to avoid any potential nonconvergence problems.

## 6.3. DC characteristics

The equivalent circuit is shown in Figure 6.1. The
parasitics include the resistances in series with all the terminals and the gate-source and the gate-drain diodes. These
are metal Schottky barrier diodes and their models were
described earlier.

Operation of the intrinsic MESFET is simliar to the
JFET and similar equations can be used:

If $v_{gs} - VT \leq 0$,

$$i_{ds} = 0$$

If $0 < v_{gs} - VT < v_{ds}$,

$$i_{ds} = \beta \left( v_{gs} - VT \right)^2 \left( 1 + \lambda v_{ds} \right)$$

If $0 < v_{ds} \leq v_{gs} - VT$,

$$i_{ds} = 2 \beta \left( 1 + \lambda v_{ds} \right) v_{ds} \left( v_{gs} - VT - \frac{1}{2} v_{ds} \right)$$

$$(6.1)$$

where,

$$v_{dsat} = v_{gs} - VT \qquad (6.2)$$

The term $1 + \lambda v_{ds}$ models the output conductance in saturation. *VT* is the threshold voltage, i.e. the value of the applied gate-source voltage, below which the channel is pinched off.

$$VT = VPO - VBI \tag{6.3}$$

where *VPO* is the pinch-off voltage determined by the channel thickness and the doping concentration and *VBI* is the built-in potential of the metal Schottky barrier which is a function of the doping concentration in the channel. The threshold voltage is also a function of the substrate bias:

$$VT = VTO + f(v_{bs}) \tag{6.4}$$

where *VTO* is the zero bias threshold voltage. Various empirical functions have been used for $f(v_{bs})$ including square root, linear and parabolic equations.

The separate equations for the linear and saturation regions can be linked by a suitable function [91] and the drain current can be written in a general form:

$$i_{ds} = i_{dsat} \tanh\left(\frac{v_{ds}}{v_{dsat}}\right)(1 + \lambda v_{ds}) \tag{6.5}$$

$$\tanh\left(\frac{v_{ds}}{v_{dsat}}\right) \approx 1 \qquad \text{for } v_{ds} > v_{dsat}$$

A suitable equation for $i_{dsat}$ is:

$$i_{dsat} = \frac{\beta\,(v_{gs} - VT\,)^n}{1 + b\,(v_{gs} - VT\,)^m} \tag{6.6}$$

*n* and *m* may be treated as user input model parameters or constant values may be used to simplify the model equations. Setting $n = 2$ and $b = 0$, gives rise to the *square law* [85] [92] characteristics in the saturation region. For some devices, the observed behavior may be significantly different from the square law relationship [93]. This can be modeled very easily by making $n$ as a user input parameter. It is observed that the square law relationship is valid for low $v_{gs} - VT$ values, whereas the behavior is linear for high $v_{gs} - VT$ [86]. This is achieved by setting $n = 2$, $m = 1$ and $b > 0$. For small $v_{gs} - VT$, the denominator is $\approx 1$ giving rise to the square law behavior. At higher $v_{gs} - VT$ however, the term $b\left(v_{gs} - VT\right) \gg 1$ resulting in a linear behavior.

The saturation voltage is given by,

$$v_{dsat} = v_{gs} - VT \qquad (6.7)$$

The actual threshold voltage is less than this value due to carrier velocity saturation effects:

$$v_{dsat} = \frac{v_{gs} - VT}{1 + \gamma \left(v_{gs} - VT\right)} \qquad (6.8)$$

For some fabrication processes and device structures, it is observed that the characteristics saturate at a constant drain-to-source voltage. In that case the following equation is used:

$$v_{dsat} = \frac{2 \text{ or } 3}{\alpha} \qquad (6.9)$$

The tanh function consumes considerable computing time and in the linear region i.e. for $x < 1$, it can be approximated by [86]:

$$\tanh(x) \approx 1 - \left(1 - \frac{x}{p}\right)^{p} \qquad p = 2 \text{ or } 3 \qquad (6.10)$$

In the saturation region $x > 1$ and $tanh(x) \approx 1$. $p = 3$ is recommended [86].

## 6.4. Subthreshold conduction

For many digital applications, it is desirable to have high levels of integration. The power dissipation in the devices must, thererfore, be kept small. Hence, devices tend to be switched between *barely on* sate and *almost off* state. Modeling of the device characteristics in the subthreshold region becomes important.

Two empirical approaches use the following equations [12] [83]:

$$I_{subth} = ISO \; e^{\left( \dfrac{(1 - \alpha \, v_{ds})(v_{gs} - VT + \gamma \, v_{ds})}{n \, v_t} \right)}$$

where,

$$v_t = \frac{k \, T}{q}$$

and $ISO$, $\alpha$, $\gamma$ and $n$ are user input model parameters.
For $v_{gs} < VTS$,

$$I_{subth} = i_{ds} \Big|_{v_{gs} = VTS}$$

$$e^{\left[ -(VTS - v_{gs}) \; \dfrac{g_m}{i_{ds}} \Big|_{v_{gs} = VTS} \right]}$$

where,

$$VTS = VT + \delta \; VPO$$

$$g_m = \frac{\partial i_{ds}}{\partial v_{gs}}$$

and $\delta$ is a user input model parameter.

## 6.5. Charge storage

Parasitic capacitances can be added between various ter minals of the MESFET. In addition to these parasitics bulk-source and bulk-drain capacitances are generally added. These capacitances can be made voltage dependent, if desired, similar to the diode capacitances.

It is observed that there is a delay between the change in gate-source voltage and a change in drain-source current. This delay can be modeled as a stored charge:

$$q_{ds} = \tau \, \dot{i}_{ds} \qquad (6.11)$$

and,

$$q_{drain} = q_{ds} \quad \text{and} \quad q_{source} = -q_{ds}$$

where $\tau$ represents the transit time under the gate. Another approach is to connect a transmission line is series with $RG$ [84] and the transmission line delay is set to $\tau$.

The remaining charge storage consists of the charges associated with the intrinsic device. The internal device capacitances were modeled earlier as capacitances of Schottky diodes connected between the gate-source and the gate-drain respectively,

$$C_{gs} = \frac{C_{gso}}{\left(1 - \dfrac{v_{gs}}{VBI}\right)^m} \qquad (6.12)$$

$$C_{gd} = \frac{C_{gdo}}{\left(1 - \dfrac{v_{gd}}{VBI}\right)^m} \qquad (6.13)$$

It is observed that such a model is completely inade-quate for gate voltages smaller than the threshold voltage. Indeed, when the channel under tha gate is pinched off, the total charge in the section of the channel under the gate does not change. But the capacitance given by these equations does not go to zero. In addition, this model does not account for the Gunn domain formation at the drain side of the gate, which is important for high pinch-off devices. Carrier velocity saturation is also not taken inot account. Due to the velocity saturation effect, $C_{gs}$ increases in the saturation region and $C_{gd}$ drops rapidly as the device enters saturation. Also, for small channel devices, the capacitance contribution due to the depletion region edges which are not directly underneath the gate, cannot be neglected.

An empirical approach which improves upon the simplified equations given above is suggested in [86]:

$$\delta = 0.2 \quad , \quad \Delta = \frac{1}{\alpha}$$

$$v_{eff2} = 0.5 \left\{ v_{gs} + v_{gd} - \sqrt{\left( v_{gs} - v_{gd} \right)^2 + \Delta^2} \right\}$$

$$v_{eff1} = 0.5 \left\{ v_{gs} + v_{gd} + \sqrt{\left( v_{gs} - v_{gd} \right)^2 + \Delta^2} \right\}$$

$$v_{eff} = 0.5 \left\{ v_{eff1} + VT + \sqrt{\left( v_{eff1} - VT \right)^2 + \delta^2} \right\}$$

$$Q_g = \int_0^{v_{eff}} C(v) \, dv + CGDN \, v_{eff2}$$

$$C(v) = \frac{CGSO}{\left( 1 - \dfrac{v}{VBI} \right)^M} \qquad \text{if } v < FC \; VBI$$

$$C(v) = \frac{CGSO}{\left(1 - FC\right)^{(1 + M)}} \left[ 1 - FC\left(1 + M\right) + \frac{v}{VBI}M \right]$$

if $v \geq FC\ VBI$

$$\Delta Q_{gs} = 0.5 \left\{ \left[ Q_g\left(v_{gs}, v_{gd}\right) - Q_g\left(v_{gs_{old}}, v_{gd}\right) \right] + \left[ Q_g\left(v_{gs}, v_{gd_{old}}\right) - Q_g\left(v_{gs_{old}}, v_{gd_{old}}\right) \right] \right\}$$

$$\Delta Q_{gd} = 0.5 \left\{ \left[ Q_g\left(v_{gs}, v_{gd}\right) - Q_g\left(v_{gs}, v_{gd_{old}}\right) \right] + \left[ Q_g\left(v_{gs_{old}}, v_{gd}\right) - Q_g\left(v_{gs_{old}}, v_{gd_{old}}\right) \right] \right\}$$

$$C_{gs} = \frac{\partial \Delta Q_{gs}}{\partial v_{gs}}$$

$$C_{gd} = \frac{\partial \Delta Q_{gd}}{\partial v_{gd}}$$

Here $CGSO$ is the zero bias gate-source capacitance and $CGDN$ is the gate-drain capacitance in saturation.

Another empirical approach is used in [83]:

$$C_{gs} = C_{GS0} + \lambda_{GS1} \left( \frac{v_{gs} - VT}{VBI - VT} \right)^{m_{GS}} f_c\left(v_{ds}\right)$$

where, for $v_{ds} \geq v_{dsat}$,

$$f_c = 1 + \lambda_{GS2} \left( \frac{v_{gs} - VT}{VBI - VT} \right)^{\frac{1}{m_{GS}}} \left( v_{ds} - v_{dsat} \right)$$

and for $v_{ds} < v_{dsat}$,

$$f_c = \left( \lambda_{GS2}\, v_{dsat} + \beta_{GS} - 1 \right) \left[ \frac{v_{ds}}{v_{dsat}} \right]^2$$

$$+ \left( 2 - \lambda_{GS2}\, v_{dsat} - 2\, \beta_{GS} \right) + \beta_{GS}$$

and,

$$C_{gd} = \frac{C_{GD1}}{\left[ 1 - \dfrac{\lambda_{GD1}\, v_{gs} - v_{ds}}{\lambda_{GD1}\, VBI} \right]^{m_{GD}}} \left( 1 - \lambda_{GD2}\, v_{gs} \right)$$

$$+ C_{GD0}$$

where $\lambda_{GS1}$, $\lambda_{GS2}$, $C_{GS0}$, $\beta_{GS}$, $m_{GS}$, $\lambda_{GD1}$, $\lambda_{GD2}$, $C_{GD0}$, $C_{GD1}$ and $m_{GD}$ are user input model parameters.

### 6.6.  Charge based model

Most of the currently available MESFET charge models use capacitance equations. As explained for MOSFET models, for accurate circuit simulations, it is necessary to analytically compute the charges associated with the device terminals. One such approach is suggested here [98] [99]. The channel charge is divided into several components, as shown in Figure 6.2. $Q_1$ and $Q_3$ are the charges associated with the edge of the depletion region at the source end. $Q_2$ and $Q_4$ are the charges associated with the edge of the depletion region at the drain end. $Q_5$ is the charge directly underneath the gate. When the device is in saturation, $Q_6$ represents the charge in the velocity saturation region of the channel and $Q_{domain}$ represents the charge associated with the domain formation. The domain formation generally occurs for high pinch-off devices, i.e. for devices having thicker channel.

Let

$a$ = channel thickness

$VT$ = threshold voltage

$VPO$ = pinch-off voltage

$VBI$ = built-in potential

$N$ = effective doping concentration in channel

$L$ = length of the device

$W$ = width of the device

$VT = VBI - VPO$

**Off region** ($v_{gs} \leq VT$):

$h_{s2} = a$

$$h_{s1} = \sqrt{\frac{2\,\epsilon}{q\,N} \left[ VBI - v_{gs} \right]}$$

$$h_{s3} = \sqrt{h_{s1}^2 - h_{s2}^2}$$

$$\tan \theta_s = \frac{h_{s2}}{h_{s3}}$$

$$Q_1 = \pi\, h_{s1}^2\, \frac{\theta_s}{2\,\pi}\, q\, N\, W$$

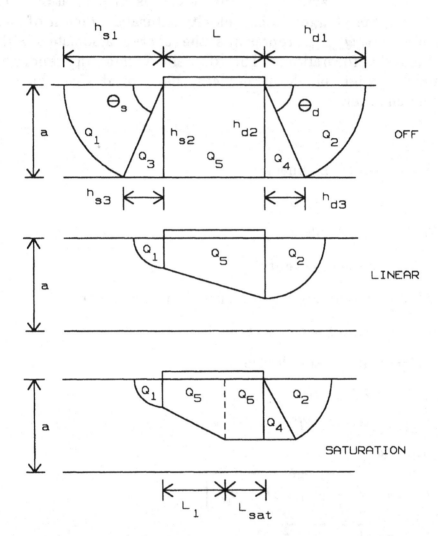

**Figure 6.2:**    MESFET charge components.

$$Q_3 = q \ N \ W \ \frac{h_{s2}}{2} \ h_{s3}$$

$$h_{d2} = a$$

$$h_{d1} = \sqrt{\frac{2\,\epsilon}{q\,N}\left[VBI - (v_{gs} - v_{ds})\right]}$$

$$h_{d3} = \sqrt{h_{d1}^2 - h_{d2}^2}$$

$$\tan\theta_d = \frac{h_{d2}}{h_{d3}}$$

$$Q_2 = \pi\,h_{d1}^2\,\frac{\theta_d}{2\,\pi}\,q\,N\,W$$

$$Q_4 = q\,N\,W\,\frac{h_{d2}}{2}\,h_{d3}$$

$$Q_5 = q\,N\,W\,L\,a$$

$$Q_{5s} = \frac{1}{2}\,Q_5$$

$$Q_{5d} = \frac{1}{2}\,Q_5$$

$$Q_s = Q_1 + Q_3 + Q_{5s}$$

$$Q_d = Q_2 + Q_4 + Q_{5d}$$

$$Q_g = -\,Q_s - Q_d$$

**Linear region** $(v_{gs} > VT,\ v_{ds} \leq v_{dsat})$:

$$h_{s3} = 0$$

$$h_{s1} = h_{s2} = \sqrt{\frac{2\,\epsilon}{q\,N}\left[VBI - v_{gs}\right]}$$

$$h_{d3} = 0$$

$$h_{d1} = h_{d2} = \sqrt{\frac{2\,\epsilon}{q\,N}\left[VBI - (v_{gs} - v_{ds})\right]}$$

$$Q_3 = Q_4 = 0$$

$$Q_1 = \frac{\pi}{4} \, h_{s1}{}^2 \, q \, N \, W$$

$$Q_2 = \frac{\pi}{4} \, h_{d1}{}^2 \, q \, N \, W$$

It is assumed that the depletion layer height varies linearly from source to drain [98]:

$$h = h_{s2} + \frac{h_{d2} - h_{s2}}{L} \, y$$

$$Q_5 = q \, N \, W \int_0^L h \, dy$$

$$= \frac{1}{2} \, q \, N \, W \, ( \, h_{d2} + h_{s2} \, ) \, L$$

$$Q_{5d} = q \, N \, W \int_0^L \frac{y}{L} \, h \, dy$$

$$= \frac{1}{2} \, q \, N \, W \left[ \, \frac{1}{3} h_{s2} + \frac{2}{3} h_{d2} \, \right] L$$

$$Q_{5s} = q \, N \, W \int_0^L \left[ 1 - \frac{y}{L} \right] h \, dy$$

$$= \frac{1}{2} \, q \, N \, W \left[ \, \frac{2}{3} h_{s2} + \frac{1}{3} h_{d2} \, \right] L$$

$$Q_s = Q_1 + Q_{5s}$$

$$Q_d = Q_2 + Q_{5d}$$

**Saturation region** ( $v_{gs} > VT$, $v_{ds} > v_{dsat}$ ):

$$h_{s3} = 0$$

$$h_{s1} = h_{s2} = \sqrt{\frac{2\,\epsilon}{q\,N}\left[VBI - v_{gs}\right]}$$

$$Q_3 = 0$$

$$Q_1 = \frac{\pi}{4}\,h_{s1}{}^2\,q\,N\,W$$

$$h_{d1} = \sqrt{\frac{2\,\epsilon}{q\,N}\left[VBI - \left(v_{gs} - v_{ds}\right)\right]}$$

$$h_{d2} = \min\left[\sqrt{\frac{2\,\epsilon}{q\,N}\left[VBI - \left(v_{gs} - v_{ds}\right)\right]},\ a\right]$$

$$h_{d3} = \sqrt{h_{d1}{}^2 - h_{d2}{}^2}$$

$$\tan\theta_d = \frac{h_{d2}}{h_{d3}}$$

$$Q_2 = \pi\,h_{d1}{}^2\,\frac{\theta_d}{2\,\pi}\,q\,N\,W$$

$$Q_4 = q\,N\,W\,\frac{h_{d2}}{2}\,h_{d3}$$

$$arg = \frac{\pi\,K_d}{2\,a}\,L\,\frac{v_{ds} - v_{dsat}}{v_{gs} - VT}\left[\frac{v_{gs} - VT}{v_{dsat}} - 1\right]$$

$$L_{sat} = \frac{2\,a}{\pi}\,\sinh^{-1}(\ arg\ )$$

$$L_1 = L - L_{sat}$$

$$Q_6 = q\,N\,W\,L_{sat}\,h_{d2}$$

$$Q_{6s} = \frac{1}{2}\,Q_6$$

$$Q_{6d} = \frac{1}{2} \, Q_6$$

$$Q_5 = q \, N \, W \int_0^{L1} h \; dy$$

$$= \frac{1}{2} \, q \, N \, W \, ( \, h_{d2} + h_{s2} \, ) \, L1$$

$$Q_{5d} = q \, N \, W \int_0^{L1} \frac{y}{L1} \, h \; dy$$

$$= \frac{1}{2} \, q \, N \, W \left[ \frac{1}{3} h_{s2} + \frac{2}{3} h_{d2} \right] L1$$

$$Q_{5s} = q \, N \, W \int_0^{L1} \left[ 1 - \frac{y}{L1} \right] h \; dy$$

$$= \frac{1}{2} \, q \, N \, W \left[ \frac{2}{3} h_{s2} + \frac{1}{3} h_{d2} \right] L1$$

$$a_{domain} = \sqrt{ \frac{2 \, \epsilon}{q \, N} \left[ VBI - v_{gs} + F_{th} \, L_1 \right] }$$

$$a_d = \min \, ( \, a_{domain} \, , \, a \, )$$

$$Q_{domain} = 0.728 \; W \left[ q \, \epsilon \, \sqrt{n_{cr} \, N} \right]^{\frac{1}{2}}$$

$$L_{sat} \left[ 1 - \frac{a_d}{a} \right] (1 - K_d)$$

$$Q_s = Q_1 + Q_{5s} + Q_{6s} + \frac{1}{2} \, Q_{domain}$$

$$Q_d = Q_2 + Q_4 + Q_{5d} + Q_{6d} + Q_{domain}$$

Some additional model parameters are introduced to model the domain formation. $F_{th}$ is the threshold field for negative differential mobility. $K_d$ is the fraction of $v_{ds} - v_{dsat}$ that is responsible for domain formation. $n_{cr}$ is the characteristic doping density (typically $3 \times 10^{15}$ $cm^{-3}$ for GaAs).

Although separate equations are given for different regions of operation for ease of explanation and understanding, it is possible to implement these equations so that various components are joined using smoothing functions across the regions.

### 6.7. Temperature dependence

Since $\beta$, the transconductance parameter, is related to the mobility, a power law relationship can be used:

$TNOM$ = nominal temperature in degrees Kelvin

$T$ = analysis temperature in degrees Kelvin

$$\beta = \beta_{TNOM} \left( \frac{T}{TNOM} \right)^{BEX}$$

where BEX is a user input model parameter, typically $-\frac{3}{2}$.

$VBI$, the built-in potential, can be computed from a functional relationship similar to the diode built-in potential. $VTO$ can then be computed from:

$$VTO = VPO - VBI$$

Alternatively, an empirical relationship may be used:

$$VTO = VTO_{NOM} \left[ 1 + TVTO \left( T - TNOM \right) \right]$$

where $TVTO$ is the first order temperature coefficient.

Since the remaining model parameters are empirical, empirical relationships can be used for these parameters, if desired.

## 6.8. Modeling for analog applications

Modeling the small signal admittances is important for analog applications. It is observed that the frequency dependence of small signal admittances computed from the large signal model equations does not agree with the dependence observed from small signal measurements [94] - [96]. Addition of $R_i$, $C_i$ at the input and $R_o$, $C_o$ at the output reduces this discrepancy. More complex analog models may be formulated by making these components voltage depedent.

## 6.9. Model parameter extraction

The extraction procedure used for parasitic resistances is different from the one used for MOSFETs. For MESFETs, since the gate forms a junction with the channel, it is possible to forward bias that junction and this property is utilized for separating $R_s$ and $R_d$. If a positive voltage is applied between the gate and the source and the drain is kept floating, as shown in Figure 6.3, then the gate current flows through half the channel resistance and the source resistance [97] [100]:

$$v_{ds} = \left( R_s + \frac{R_{ch}}{2} \right) i_g$$

The slope of the $v_{ds}$ versus $i_g$ curve gives $R_s + \dfrac{R_{ch}}{2}$. Now, a similar measurement with drain grounded and source floating will give $R_d + \dfrac{R_{ch}}{2}$. If a small bias is applied to the drain in Figure 6.3, so that $i_{ds} \ll i_g$, then the drain current will not substantially change the potential distribution in the channel and we can write:

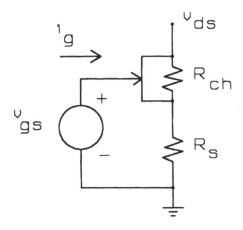

**Figure 6.3:** "End" resistance measurement technique.

$$v_{ds} = \left(R_s + R_d + R_{ch}\right) i_{ds} + \left(R_s + \frac{R_{ch}}{2}\right) i_g$$

Hence, $R_s + R_d + R_{ch}$ can be found from the intercept of the $v_{ds}$ versus $i_g$ curve. The three equations can then be solved to obtain $R_{ch}$, $R_s$ and $R_d$.

Once the parasitic resistances are known, the other model parameters can be found using techniques similar to the MOSFETs. Good initial estimates are used in conjunction with an optimization program to match the measured device characteristics.

# REFERENCES

[1]    Ian Getreu, Modeling The Bipolar Transistor, Tek-
       tronix Inc., Beaverton, Oregon, 1976.

[2]    Leon O. Chua and Pen-Min Lin, Computer-Aided
       Analysis of Electronic Circuits, Prentice-Hall Inc.,
       1975.

[3]    L.W.Nagel, "SPICE2: A Computer Program to Simu-
       late Semiconductor Circuits", Memorandum No.
       ERL-M520, Electronics Research Laboratory, College
       of Engineering, University of California, Berkeley,
       CA 94720, May 1975.

[4]    W.J.McCalla    and    D.O.Pederson,    "Elements    of
       Computer-Aided Circuit Analysis", IEEE Trans. on
       Circuit Theory, Vol. CT-18, No. 1, pp. 14 - 26, Janu-
       ary 1971.

[5]    A.E.Ruehli and G.S.Ditlow, "Circuit Analysis, Logic
       Simulation, and Design Verification for VLSI", Proc.
       of the IEEE, Vol. 71, No. 1, pp. 34 - 48, January
       1983.

[6]    P.Yang, B.D.Epler and P.K.Chatterjee, "An Investiga-
       tion of the Charge Conservation Problem for MOS-
       FET Circuit Simulation", IEEE Journal of Solid-State
       Circuits, Vol. SC-18, No. 1, pp. 128 - 138, February
       1983.

[7]    D.E.Ward and K.Doganis, "Optimized Extraction of
       MOS Model Parameters", IEEE Trans. on CAD, Vol.
       CAD-1, No. 4, pp. 163 - 168, October 1982.

[8]    K.Doganis and D.L.Scharfetter, "General Optimiza-
       tion and Extraction of IC Device Model Parameters",
       IEEE Trans. on Electron Devices, Vol. ED-30, No. 9,
       pp. 1219 - 1228, September 1983.

[9]    P.Yang and P.K.Chatterjee, "An Optimal Parameter
       Extraction Program for MOSFET Models", IEEE
       Trans. on Electron Devices, Vol. ED-30, No. 9, pp.
       1214 - 1219, September 1983.

[10]   R.J.Sokel and D.B.Macmillen, "Practical Integration
       of Process, Device, and Circuit Simulation", IEEE
       Trans. on CAD, Vol. CAD-4, No. 4, pp. 554 - 560,
       October 1985.

[11]   D.W.Marquardt, "An Algorithm for Least-Squares
       Estimates of Non-Linear Parameters", SIAM Journal,
       Vol. 11, No. 2, pp. 431 - 441, June 1963.

[12]   C.T.M.Chang, T.Vrotsos, M.T.Frizzell and R.Carroll,
       "A    Subthreshold    Current    Model    for    GaAs
       MESFET's", IEEE Electron Device Letters, Vol.
       EDL-8, No. 2, pp. 69 - 72, February 1987.

[13]   D.B.Estreich, "A Simulation Model for Schottky
       Diodes in GaAs Integrated Circuits", IEEE Trans. on

CAD, Vol. CAD-2, No. 2, pp. 106 - 111, April 1983.

[14]   W.J.McCalla, Computer-Aided Circuit Simulation Techniques, Kluwer Academic Publishers, 1987.

[15]   S.M.Sze, Physics of Semiconductor Devices, John Wiley & Sons, 1981.

[16]   A.S.Grove, Physics and Technology of Semiconductor Devices, John Wiley & Sons, 1967.

[17]   J.G.J.Chern, P.Chang, R.F.Motta and N.Godinho, "A New Method to Determine MOSFET Channel Length", IEEE Electron Device Letters, Vol. EDL-1, No. 9, pp. 170 - 173, September 1980.

[18]   Y.R.Ma and K.L.Wang, "A New Method to Electrically Determine Effective MOSFET Channel Width", IEEE Trans. on Electron Devices, Vol. ED-29, No. 12, pp. 1825 - 1827, December 1982.

[19]   F.H.DeLaMoneda, H.N.Kotecha and M.Shatzkes, "Measurement of MOSFET Constants", IEEE Electron Device Letters, Vol. EDL-3, No. 1, pp. 10 - 12, January 1982.

[20]   M.J.Thoma and C.R.Westgate, "A New AC Measurement Technique to Accurately Determine MOSFET Constants", IEEE Trans. on Electron Devices, Vol. ED-31, No. 9, pp. 1113 - 1116, September 1984.

[21]   S.E.Laux, "Accuracy of an Effective Channel Length / External Resistance Extraction Algorithm for MOSFET's", IEEE Trans. on Electron Devices, Vol. ED-31, No. 9, pp. 1245 - 1251, September 1984.

[22]   B.J.Sheu, C.Hu, P.K.Ko and F.C.Hsu, "Source-and-Drain Series Resistance of LDD MOSFET's", IEEE Electron Device Letters, Vol. EDL-5, No. 9, pp. 365 - 367, September 1984.

[23]   K.L.Peng, S.Y.Oh, M.A.Afromowitz and J.L.Moll, "Basic Parameter Measurement and Channel

Broadening Effect in the Submicrometer MOSFET",
IEEE Electron Device Letters, Vol. EDL-5, No. 11,
pp. 473 - 475, November 1984.

[24] M.H.Seavey, "Source and Drain Resistance Determination for MOSFET's", IEEE Electron Device
Letters, Vol. EDL-5, No. 11, pp. 479 - 481,
November 1984.

[25] B.J.Sheu and P.K.Ko, "A Simple Method to Determine Channel Widths for Conventional and LDD
MOSFET's", IEEE Electron Device Letters, Vol.
EDL-5, No. 11, pp. 485 - 486, November 1984.

[26] B.J.Sheu and P.K.Ko, "A Capacitance Method to
Determine Channel Lengths for Conventional and
LDD MOSFET's", IEEE Electron Device Letters,
Vol. EDL-5, No. 11, pp. 491 - 493, November 1984.

[27] J.Whitfield, "A Modification on "An Improved
Method to Determine MOSFET Channel Length"",
IEEE Electron Device Letters, Vol. EDL-6, No. 3, pp.
109 - 110, March 1985.

[28] J.Scarpulla and J.P.Krusius, "Improved Statistical
Method for Extraction of MOSFET Effective Channel
Length and Resistance", IEEE Trans. on Electron
Devices, Vol. ED-34, No. 6, pp. 1354 - 1359, June
1987.

[29] H.Shicman and D.A.Hodges, "Modeling and Simulation of Insulated-Gate Field-Effect Transistor Switching Circuits", IEEE Journal of Solid-State Circuits,
Vol. SC-3, No. 5, pp. 285 - 289, September 1968.

[30] C.Duvvury, "A Guide to Short-Channel Effects in
MOSFETs", IEEE Circuits and Devices Magazine, pp.
6 - 10, November 1986.

[31] H.I.Hanafi, L.H.Camintz and A.J.Dally, "An Accurate
and Simple MOSFET Model for Computer-Aided

Design'', IEEE Journal of Solid-State Circuits, Vol. SC-17, No. 5, pp. 882 - 891, October 1982.

[32]  P.P.Wang, ''Device Characteristics of Short-Channel and Narrow-Width MOSFET's'', IEEE Trans. on Electron Devices, Vol. ED-25, No. 7, pp. 779 - 786, July 1978.

[33]  VLSI LABORATORY, Texas Instruments Inc., ''Technology and Design Challenges of MOS VLSI'', IEEE Journal of Solid-State Circuits, Vol. SC-17, No. 3, pp. 442 - 448, June 1982.

[34]  L.D.Yau, ''A Simple Theory to Predict the Threshold Voltage of Short-Channel IGFET's'', Solid State Electronics, Vol. 17, pp. 1059 - 1063, 1974.

[35]  G.Merckel, ''A Simple Model of the Treshold Voltage of Short and Narrow Channel MOSFETs'', Solid State Electronics, Vol. 23, pp. 1207 - 1213, 1980.

[36]  W.L.Engl, H.K.Dirks and B.Meinerzhagen, ''Device Modeling'', IEEE Proceedings, Vol. 71, No. 1, pp. 10 - 33, January 1983.

[37]  D.Frohman-Bentchkowsky and A.S.Grove, ''Conductance of MOS Transistors in Saturation'', IEEE Trans. on Electron Devices, Vol. ED-16, No. 1, pp. 108 - 113, January 1969.

[38]  R.R.Troutman and S.N.Chakravarti, ''Subthreshold Characteristics of Insulated-Gate Field-Effect Transistors'', IEEE Trans. on Circuit Theory, Vol. CT-20, No. 6, pp. 659 - 665, November 1973.

[39]  R.M.Swanson and J.D.Meindl, ''Ion-Implanted Complementary MOS Transistors in Low-Voltage Circuits'', IEEE Journal of Solid-State Circuits, Vol. SC-7, No. 2, pp. 146 - 153, April 1972.

[40]  P.Antognetti, D.D.Caviglia and E.Profumo, ''CAD Model for Threshold and Subthreshold Conduction in

MOSFET's'', IEEE Journal of Solid-State Circuits, Vol. SC-17, No. 3, pp. 454 - 458, June 1982.

[41]    P.C.Chan, R.Liu, S.K.Lau and M.Pinto-Guedes, ''A Subthreshold Conduction Model for Circuit Simulation of Submicron MOSFET'', IEEE Trans. on CAD, Vol. CAD-6, No. 4, pp. 574 - 581, July 1987.

[42]    J.Mar, S.Li and S.Yu, ''Substrate Current Modeling for Circuit Simulation'', IEEE Trans. on CAD, Vol. CAD-1, No. 4, pp. 183 - 186, October 1982.

[43]    T.Y.Chan, P.K.Ko and C.Hu, ''A Simple Method to Characterize Substrate Current in MOSFET's'', IEEE Electron Device Letters, Vol. EDL-5, No. 12, pp. 505 - 507, December 1984.

[44]    J.J.Barnes, K.Shimohigashi and R.W.Dutton, ''Short-Channel MOSFET's in the Punchthrough Current Mode'', IEEE Trans. on Electron Devices, Vol. ED-26, No. 4, pp. 446 - 453, April 1979.

[45]    R.B.Fair and R.C.Sun, ''Threshold-Voltage Instability in MOSFET's Due to Channel Hot-Hole Emission'', IEEE Trans. on Electron Devices, Vol. ED-28, No. 1, pp. 83 - 94, January 1981.

[46]    G.T.Wright and H.M.A. Gaffur, ''Preprocessor Modeling of Parameter and Geometry Dependences of Short and Narrow MOSFET's for VLSI Circuit Simulation, Optimization, and Statistics with SPICE'', IEEE Trans. on ELectron Devices, Vol. ED-32, No. 7, pp. 1240 - 1245, July 1985.

[47]    M.C.Hsu and B.J.Sheu, ''Inverse-Geometry Dependence of MOS Transistor Electrical Parameters'', IEEE Trans. on CAD, Vol. CAD-6, No. 4, pp. 582 - 585, July 1987.

[48]    A.Vladimirescu and S.Liu, ''The Simulation of MOS Integrated Circuits Using SPICE2'', Memorandum

No. UCB/ERL M80/7, Electronics Research Laboratory, University of California, Berkeley, CA 94720, February 1980.

[49] S.L.Wong and C.A.T.Salama, "Improved Simulation of p- and n-channel MOSFET's Using an Enhanced SPICE MOS3 Model", IEEE Trans. on CAD, Vol. CAD-6, pp. 586 - 591, July 1987.

[50] S.Liu and L.W.Nagel, "Small-Signal MOSFET Models for Analog Circuit Design", IEEE Journal of Solid-State Circuits, Vol. SC-17, No. 6, pp. 983 - 998, December 1982.

[51] B.J.Sheu, D.L.Scharfetter and H.C.Poon, "Compact Short-Channel IGFET Model (CSIM)", Memorandum No. UCB/ERL M84/20, Electronics Research Laboratory, University of California, Berkeley, CA 94720, March 1984.

[52] B.J.Sheu, D.L.Scharfetter, P.K.Ko and M.C.Jeng, "BSIM: Berkeley Short-Channel IGFET Model for MOS Transistors", IEEE Journal of Solid-State Circuits, Vol. SC-22, No. 4, pp. 558 - 566, August 1987.

[53] A.R.Newton, "Timing, Logic and Mixed-Mode Simulation for Large MOS Integrated Circuits", Computer Design Aids for VLSI Circuits, Sijthoff & Noordhoff International Publishers, pp. 175 - 239, The Hague, 1981.

[54] T.Shima, T.Sugawara, S.Moriyama and H.Yamada, "Three-Dimensional Table Look-up MOSFET Model for Precise Circuit Simulation", IEEE Journal of Solid-State Circuits, Vol. SC-17, No. 3, pp. 449 - 454, June 1982.

[55] N.K.Jain, D.Agnew and M.S.Nakhla, "Two-Dimensional Table Look-up MOSFET Model", IEEE International Conference on Computer-Aided Design,

pp. 201 -203, 1983.

[56]   T.Shima, H.Yamada and R.L.M.Dang, "Table Look-up MOSFET Modeling System Using a 2-D Device Simulator and Monotonic Piecewise Cubic Interpolation", IEEE Trans. on CAD, Vol. CAD-2, No. 2, pp. 121 - 126, April 1983.

[57]   J.L.Burns, A.R.Newton and D.O.Pederson, "Active Device Table Look-up Models for Circuit Simulation", International Symposium on Circuits and Systems, pp. 250 - 253, 1983.

[58]   P.Subramaniam, "Table Models for Timing Simulation", IEEE Custom Integrated Circuits Conference, pp. 310 - 314, 1984.

[59]   G.Bischoff and J.P.Krusius, "Technology Independent Device Modeling for Simulation of Integrated Circuits for FET Technologies", IEEE Trans. on CAD, Vol. CAD-4, No. 1, pp. 99 - 110, January 1985.

[60]   K.Sakui, T.Shima, Y.Hayashi, F.Horiguchi and M.Ogura, "A Simplified Accurate Three-dimensional Table Look-up MOSFET Model for VLSI Circuit Simulation", IEEE Custom Integrated Circuits Conference, pp. 347 - 351, 1985.

[61]   D.Divekar, D.Ryan, J.Chan and J.Deutsch, "Fast and Accurate Table Look-up MOSFET Model for Circuit Simulation", IEEE Custom Integrated Circuits Conference, pp. 621 - 623, 1986.

[62]   T.Shima, "Table Lookup MOSFET Capacitance Model for Short-Channel Devices", IEEE Trans. on CAD, Vol. CAD-5, No. 4, pp. 624 - 632, October 1986.

[63]   Y.A.El-Mansy, "Analysis and Characterization of the Depletion-mode IGFET", IEEE. Journal of Solid-State Circuits, Vol. Sc-15, No. 3, pp. 331 - 340, June

1980.

[64] D.A.Divekar and R.I.Dowell, "A Depletion-Mode MOSFET Model for Circuit Simulation", IEEE Trans. on CAD, Vol.CAD-3, No. 1, pp. 80 - 87, January 1984.

[65] K.C.K.Weng, P.Yang and J.H.Chern, "A Predictor/CAD Model for Buried-Channel MOS Transistors", IEEE Trans. on CAD, Vol. CAD-6, No. 1, pp. 4 - 16, January 1987.

[66] J.E.Meyer, "MOS Models and Circuit Simulation", RCA Review, Vol. 32, pp. 42 - 63, March 1971.

[67] T.K.Young and R.W.Dutton, "MINI-MSINC - A Minicomputer Simulator for MOS Circuits with Modular Built-in Model", Technical Report No. 5013-1, Stanford Electronics Laboratories, Stanford University, Stanford, CA 94305, March 1976.

[68] D.E.Ward and R.W.Dutton, "A Charge-Oriented Model for MOS Transistor Capacitances", IEEE Journal of Solid-State Circuits, Vol. SC-13, No. 5, pp. 703 - 708, October 1978.

[69] D.E.Ward, "Charge-Based Modeling of Capacitance in MOS Transistors", Technical Report No. G201-11, Stanford Electronics Laboratories, Stanford University, Stanford, CA 94305, June 1981.

[70] P.Mancini, C.Turchetti and G.Masetti, "A Non-Quasi-Static Analysis of the Transient Behavior of the Long-Channel MOST Valid in All Regions of Operation", IEEE Trans. on Electron Devices, Vol. ED-34, No. 2, pp. 325 - 334, February 1987.

[71] J.J.Paulos, D.A.Antoniadis and Y.P.Tsividis, "Measurement of Intrinsic Capacitances of MOS Transistors", IEEE International Solid-State Circuits Conference, pp. 238 - 239, February 1982.

[72] J.Oristian, H.Iwai, J.Walker and R.Dutton, "Small Geometry MOS Transistor Capacitance Measurement Method Using Simple On-Chip Circuits", IEEE Electron Device Letters, Vol. EDL-5, No. 10, pp. 395 - 397, October 1984.

[73] B.J.Sheu and P.K.Ko, "Short-Channel Effects on MOS Transistor Capacitances", IEEE Trans. on Circuits and Systems, Vol. CAS-33, No. 10, pp. 1030 - 1032, October 1986.

[74] H.Iwai, M.R.Pinto, C.S.Rafferty, J.E.Oristian and R.W.Dutton, "Analysis of Velocity Saturation and Other Effects on Short-Channel MOS Transistor Capacitances", IEEE Trans. on CAD, Vol. CAD-6, No. 2, pp. 173 - 184, March 1987.

[75] Y.Tsividis, "Moderate Inversion in MOS Devices", Solid-State Electronics, Vol. 25, No. 11, pp. 1099 - 1104, 1982.

[76] Y.Tsividis and G.Masetti, "Problems in Precision Modeling of the MOS Transistor for Analog Applications", IEEE Trans. on CAD, Vol. CAD-3, No. 1, pp. 72 - 79, January 1984.

[77] J.R.Brews, "A Charge Sheet Model of the MOSFET", Solid-State Electronics, Vol. 21, pp. 345 - 355, 1978.

[78] F.Van de Wiele, "A Long-Channel MOSFET Model", Solid-State Electronics, Vol. 22, pp. 991 - 997, 1979.

[79] H.A.Nienhaus, J.C.Bowers and P.C.Herren, "A High Power MOSFET Computer Model", Powerconversion International, pp. 65 - 73, January 1982.

[80] H.Cheng and A.G.Milnes, "Power MOSFET Characteristics with Modified SPICE Modeling", Solid-State Electronics, Vol. 25, No. 12, pp. 1209 - 1212, 1982.

[81]    R.C.Eden,      A.R.Livingston      and      B.M.Welch,
        "Integrated    Circuits:    The    Case    For    Gallium
        Arsenide", IEEE Spectrum, pp. 30 - 37, December
        1983.

[82]    S.E.Sussman-Fort, S.Narasimhan and K.Mayaram, "A
        Complete GaAs MESFET Computer Model for
        SPICE", IEEE Trans. on Microwave Theory and
        Techniques, Vol. MTT-32, No. 4, pp. 471 - 473, April
        1984.

[83]    J.M.Golio, J.R.Hauser and P.A.Blakey, "A Large-
        Signal GaAs MESFET Model Implemented on
        SPICE", IEEE Circuits and Devices Magazine, pp. 21
        - 30, September 1985.

[84]    C.Huang and A.R.Thorbjornsen, "A SPICE Modeling
        Technique for GaAs MESFET IC's", IEEE Trans. on
        Electron Devices, Vol. ED-32, No. 5, pp. 996 - 998,
        May 1985.

[85]    C.H.Hyun, M.S.Shur and A.Peczalski, "Analysis of
        Noise Margin and Speed of GaAs MESFET DCFL
        Using UM-SPICE", IEEE Trans. in Electron Devices,
        Vol. ED-33, No. 10, pp. 1421 - 1426, October 1986.

[86]    H.Statz, P.Newman, I.W.Smith, R.A.Pucel and
        H.A.Haus, "GaAs FET Device and Circuit Simulation
        in SPICE", IEEE Trans. on Electron Devices, Vol.
        ED-34, No. 2, pp. 160 - 169, February 1987.

[87]    A.Peczalski, C.H.Chen, M.S.Shur and S.M.Baier,
        "Modeling and Characterization of Ion-Implanted
        GaAs MESFET's", IEEE Trans. on Electron Devices,
        Vol. ED-34, No. 4, pp. 726 - 732, April 1987.

[88]    R.Goyal and N. Scheinberg, "GaAs MESFET Model
        for Precision Analog IC Design", VLSI Systems
        Design, pp. 52 - 55, May 4, 1987.

[89]    S.G.Peltan, S.I.Long and S.E.Butner, "An Accurate

DC SPICE Model for the GaAs MESFET'', IEEE International Symposium on Circuits and Systems, pp. 6 - 11, May 1987.

[90] S.E.Sussman-Fort and J.C.Hantgan, ''A SPICE-Installed Model for Enhancement and Depletion Mode GaAs FETs'', IEEE International Symposium on Circuits and Systems, pp. 15 - 18, May 1987.

[91] T.Taki, ''Approximation of Junction Field-Effect Transistor Characteristics by a Hyperbolic Function'', IEEE Journal of Solid-State Circuits, Vol. SC-13, No. 5, pp. 724 - 726, October 1978.

[92] W.R.Curtice, ''A MESFET Model for Use in the Design of GaAs Integrated Circuits'', IEEE Trans. on Microwave Theory and Techniques, Vol. MTT-28, No. 5, pp. 448 - 456, May 1980.

[93] L.E.Larson, ''An Improved GaAs MESFET Equivalent Circuit Model for Analog Integrated Circuit Applications'', IEEE Journal of Solid-State Circuits, Vol. SC-22, No. 4, pp. 567 - 574, August 1987.

[94] W.R.Curtice and R.L.Camisa, ''Self-Consistent GaAs FET Models for Amplifier Design and Device Diagnostics'', IEEE Trans. on Microwave Theory and Techniques, Vol. MTT-32, No. 12, pp. 1573 - 1578, December 1984.

[95] N.Scheinberg, ''Designing High Speed Operational Amplifiers with GaAs MESFETs'', IEEE International Symposium on Circuits and Systems, pp. 193 - 198, May 1987.

[96] L.E.Larson, ''Gallium Arsenide MESFET Modeling for Analog Integrated Circuit Design'', IEEE International Symposium on Circuits and Systems, pp. 1 - 5, May 1987.

[97] K.W.Lee, K.Lee, M.S.Shur, T.T.Vu, P.C.T.Roberts

and M.J.Helix, "Source, Drain, and Gate Series Resistances and Electron Saturation Velocity in Ion-Implanted GaAs FET's", IEEE Trans. on Electron Devices, Vol. ED-32, No.5, pp. 987 - 992, May 1985.

[98] T.H.Chen and M.S.Shur, "A Capacitance Model for GaAs MESFET's", IEEE Trans. on Electron Devices, Vol. ED-12, No. 5, pp. 883 - 891, May 1985.

[99] T.Takada, K.Yokoyama, M.Ida and T.Sudo, "A MES-FET Variable-Capacitance Model for GaAs Integrated Circuit Simulation", IEEE Trans. on Microwave Theory and Techniques, Vol. MTT-30, No. 5, pp. 719 - 733, May 1982.

[100] K.Lee, M.Shur, K.W.Lee, T.Vu, P.Roberts and M.Helix, "A New Interpretation of "End" Resistance Measurements", IEEE Electron Device Letters, Vol. EDL-5, No. 1, pp. 5 - 7, January 1984.

# INDEX